Celestial Mirror

Celestial Mirror

The Astronomical Observatories of Jai Singh II

Barry Perlus

Yale
UNIVERSITY PRESS

New Haven and London

Published with assistance from the Annie Burr Lewis Fund and from the foundation established in memory of Amasa Stone Mather of the class of 1907, Yale College.

Yale University Press books may be purchased in quantity for educational, business, or promotional use. For information, please e-mail sales.press@yale.edu (U.S. office) or sales@yaleup.co.uk (U.K. office).

Designed by Barry Perlus.

Printed in Hong Kong.

Library of Congress Control Number: 2019950192

ISBN 978-0-300-24627-8 (hardcover : alk. paper)

A catalogue record for this book is available from the British Library.

This paper meets the requirements of ANSI/NISO Z39.48-1992 (Permanence of Paper).

10 9 8 7 6 5 4 3 2 1

CONTENTS

Introduction

I first encountered the Jantar Mantars, as Jai Singh's observatories have come to be known, in 1989. I was so deeply impressed that I returned several times in the decades that followed to photograph and learn about them. Now, nearly 300 years after they were created, the observatories retain a powerful presence, conveying silently through their forms the beauty and mystery of our relationship with the heavens.

This book celebrates the Jantar Mantars through combinations of text and image designed to both inform and inspire. It features a section explaining the astronomy and design Jai Singh used in his observatories, and a portfolio of panoramic photographs presented as full-page spreads. Anisha Shekhar Mukherji's introductory essay "Time and Space in the Jantar Mantars" gives an in-depth explanation of the principles of Indian astronomy and philosophy, without which any understanding of this extraordinary architecture would be incomplete.

The book combines photography and research with my current interest in immersive imaging. As such, it comprises two main themes, each with its own section: *Explanation* and *Immersion.*

The explanation section includes site plans, illustrations, diagrams, and photographs addressing questions about how naked-eye astronomy works and the principles underlying Jai Singh's designs. It also provides illustrations and descriptions of the major instruments at each of the observatories. While an understanding of trigonometry and a familiarity with astronomical concepts are helpful in understanding how the instruments worked, it is not necessary to have this background to appreciate the beauty of Jai Singh's designs.

The immersion section presents full-page photographs derived from the panoramas I made during visits in 2001 and 2004. While immersive imaging is not a new phenomenon (precedents date back to as early as 60 B.C.), it is still not widely appreciated or understood.

My interest in presenting this work stems from my ongoing exploration of wide-field and panoramic photography, and the excitement I felt when I saw my panoramas projected in the immersive dome theater of Chicago's Adler Planetarium. For Jai Singh's observatories are not just for looking at. The power of his vision is best felt by being there, standing within the architecture he created to measure the sky. And although the book is a form that we usually think of looking *at*, it is my hope that this book will encourage you to see differently and, with a little imagination, experience what it is like to be within one of the observatories.

Jantar Mantar

To anyone not yet familiar with Jai Singh's observatories, the name *Jantar Mantar* has a distinctly eastern feel, while conjuring other rhyming pairs like hocus pocus or topsy turvy. Although Jai Singh never called his observatories by this name (he referred to them as *Jantras*), it has come to represent them for the past 200 years. Scholars suggest that the name came about as a colloquial version of Jai Singh's term *jantra* combined with the term *mantra* (magic words), perhaps based on the sight of astronomers bent over the instruments' scales taking measurements in what might seem a ritualistic or magical process.

Jai Singh

Maharaja Sawai Jai Singh II was born in 1688 in the town of Amber, Rajasthan. The young Jai Singh was intelligent, resourceful, and showed an early interest in mathematics and astronomy. As a prince in the royal Kacchawaha family, he would have been given ample opportunity to study a range of subjects to develop the skills and intellectual breadth for leadership. With the death of his father, Bisan Singh, in 1700, Jai Singh became king and began to assume his responsibilities as a Rajput serving the Mughal Emperor Aurangzeb.

The Mughal Empire (1526-1857) at its height extended from Kabul, Afghanistan, in the west, to Chittagong, Bangladesh, in the east, and from Kashmir in the north to the Deccan Plateau in the south. Between the reigns of Akbar and Shah Jahan (1556-1658) the empire supported a flourishing economy with worldly trade and exchange, tolerance for diverse religions, and support of the arts and sciences, including knowledge exchanges between European, Islamic, and Hindu scientists and scholars. It was in this context that Jai Singh's studies thrived,

By the time Jai Singh ascended the throne, the empire had begun to decline, and with Aurangzeb's death in 1707 there began a period of political instability that lasted until Muhammad Shah became emperor in 1719. During those turbulent times, Jai Singh managed to preserve sovereignty over his ancestral lands while continuing to advance his studies. It was during this time that he met pandit Jagannatha Samrat, a young Brahmin from Maharashtra, who joined the Maharaja's court and became a trusted guru and advisor. Jagannatha later played an important role as royal astronomer and collaborator in the design of the Jantar Mantar instruments. Although Jai Singh was aware of earlier observatories built by Greek and Persian astronomers, and based some of his designs on them, his instruments were more sophisticated and often performed multiple functions. A number of instruments designed by Jai Singh have no precedent.

Under Muhammad Shah, Jai Singh found political recognition, and was given a formal title and governorship over important Mughal provinces. It is not surprising then, that the first observatory he completed was built with the emperor's approval in Jaisinghpura, an area of vacant land owned by Jai Singh that was in close proximity to the emperor's seat of power in Delhi. The observatory was completed in 1724, and between 1724 and 1738 Jai Singh built four additional observatories at Mathura, Ujjain, Varanasi, and Jaipur. Of all the observatories, the most elaborate and largest is the observatory in Jaipur, built just outside the grounds of the royal palace. Although we think of Jai Singh as a monarch and astronomer, he was responsible for many other projects, including the city that bears his name. As the late architectural historian Bonnie G. MacDougall tells us:

> His largest project was the design and building of Jaipur (c. 1727), which has served as a regional center from the eighteenth century until the present day. The construction of a new town on such a scale was unprecedented in the annals of Indian town planning. Its design, especially its east-west ceremonial axis and its rectilinear grid pattern, was also remarkable. Formally and conceptually, Jaipur was without parallel in India until the twentieth century.

Jai Singh died in 1743 at the age of 54, only a short time after the completion of the Jaipur observatory.

www.jantarmantar.org

For more ways to explore Jai Singh's observatories, visit *www.jantarmantar.org*. It's the website I created to present the observatories to a wider audience and is the perfect complement to this book.

Time and Space in the Jantar Mantars

Anisha Shekhar Mukherji

The Jantar Mantars cannot be described easily.

Their extraordinary forms defy any categorization. Confronted with these strikingly different buildings, visitors often fall back on their own references to make sense of them. So, to a British officer in the nineteenth century, they seem to have been "manufactured by the Titans, in order to take a reconnaissance of the heavens, before they commenced their siege of Olympus."[1] "The paradise of an early Cubist"[2] is how some architects see them. And to a tourist in this century, they are just "a collection of weird buildings contorted beyond belief."[3]

Though they have drawn much attention as well as inspired a range of varied interpretations over the years, this has generally been limited to their appearance, at the cost of the very reasons that generated the creation of the Jantar Mantars. There are important exceptions of course, such as Professor V. N. Sharma's detailed field research and analysis of the principles in *Sawai Jai Singh and His Observatories*.[4] Or the explanation of the instruments in *Stone Observatories in India*,[5] by Prahlad Singh, assisted by Pandit Kalyan Dutt Sharma, Jyotishacharya[6] and former supervisor of the Jaipur Jantar Mantar. Originally termed "Jantra,"[7] these observatories at Delhi, Jaipur, Ujjain, Benaras, and Mathura have been variously called

The unusual forms of the Jantar Mantar

Yantra Mandir, *Yantra Mantra,* and *Jantar Mantar*. Almost all of them along with their original records have been vandalized, destroyed, or reinterpreted to a greater or lesser extent down the centuries—especially after the colonization of the Indian subcontinent. Consequently, most available accounts about the Jantar Mantars are by European travelers, merchants, soldiers, and civil servants, who descended on India in large numbers from the eighteenth century onward. Their accounts are inevitably colored by a belief in the superiority of their own race, culture, and religion. As an instance, William Hunter, who wrote about the Jantar Mantars in comparative detail after visiting them in the late eighteenth century, confides in an article published in *Asiatic Researches*:

> I have always thought, that after having convinced the Eastern nations of our superiority in policy and in arms, nothing can contribute more to the extension of our national glory, than a diffusion among them of a taste for European science.[8]

Thus, even among those who do recognize the connection of the Jantar Mantars with the science of astronomy, how they are so connected is insufficiently investigated.[9] Information in the public realm about how, when, and why they were built is scarce and often contradictory. Many writers claim that they are nothing more than copies of the instruments at the fifteenth-century Samarkand observatory of Mirza Ulugh Beg or that they are completely derived from Persian and Greek schools of astronomy; others write that they are developments of the Hindu school of astronomy. And so it goes on.

The intent of clarifying some of these issues, along with related aspects of the socio-political conditions that effected the establishment of this set of observatories, gave impetus to the writing of my book *Jantar Mantar, Sawai Jai Singh's Observatory in Delhi*. This, as the earliest of the Jantar Mantars, is a site of great cultural significance, and one that I have had the opportunity to conduct research on and observe closely as part of my long association with formulating its conservation policy.

It is with this background that the publication of *Celestial Mirrors* assumes importance for me. First, of course, because the photographic technology used in it, as well as the empathy of the author, makes the experience of seeing the Jantar Mantars different from their representations in other books. And second, because—despite playing up their unusual physical presence to immerse the viewer—this book recognizes and underlines the practical objective of the Jantar Mantars. It presents a concise survey of the main instruments in all the four extant Jantar Mantars, along with appropriate visual information including diagrams about their working. That these are accompanied by a lucid explanation of general principles of astronomy furthers the exploration of these complex built-structures, while literally giving readers a different perspective in experiencing their dramatic spaces.

Clarifying the Context

However, apart from understanding the specific functions of the various yantra or instruments in the Jantar Mantars, we need to comprehend the world-view that generated their formulation. Many of us—even those who are heirs to this world-view—owing to the current mode of looking at science and education from an exclusively modernist vantage-point, cannot conceive of any alternative to this mode. So it is that we miss the fact that despite being so singular, the Jantar Mantars are an intrinsic part of an indigenous Indian tradition of knowledge-systems, which in turn spring from a particular conception of the universe. It is this context of the Jantar Mantars, one rarely acknowledged, that I propose to clarify.

Indeed, the Indian world-view is specifically characterized by the primacy accorded to context. This is an extension of the objective of an ideal aesthetic, spiritual and ecological balance, often termed *dharma*.[10] The word *dharma* is derived from the root *dhr*: to uphold, sustain, or nourish.[11] In this world-view, far more emphasis is given to **why** an event has occurred or a text or building made; **how** this does or does not attain the balance of dharma; and therefore **why** it should be remembered—rather than when and where it occurred.

Instead of seeing them merely as a collection of fantastic shapes or archaeological remains or even just as instruments of astronomy, the Jantar Mantars have to be seen as a response to, as well as a manifestation of, this striving for an ideal context. All the questions that arise in our minds regarding these enigmatic structures are of value only inasmuch as they clarify the "why" of these observatories: in other words, their conception within the larger philosophical framework of the Indian culture.

At a fundamental level, the Jantar Mantars express concepts of *kala* and *aakash,* time and space, in the Indian tradition. This tradition has always recognized the fact that these are mental constructs; that they pervade all creation and all things; that they are unknowable in their entirety—and yet it has also devised very accurate methods for their measurement. The *Upanishad*s, whixh encompass the essence of the Indian world-view, put it thus: "From time all beings flow, from time they advance to growth, in time they obtain rest. Time is formed and formless."[12] And the *Surya Siddhantha,* a work on astronomy deemed to be of great antiquity and foremost importance in the Indian tradition, describes "forms of time of invisible shape," and goes on to list nine different modes of reckoning time while clarifying that practical use is made of only four modes—solar, lunar, sidereal, and civil.[13]

These modes of time are connected at the most basic plane in the form of *prana,* the spirit which literally sustains us as the "breath" we take in the present moment. Different measures are based on this unit of prana to comprehend and communicate scales that are many times smaller and larger than the normal human experience—in a system akin to that used by scientists and astronomers today. For instance, one prana, or the time taken for a full inhalation (equivalent to 4 seconds), is sub-divided further into instants of time that are as minute as 1/600 of a prana (1/150 of a second).[14] Simultaneously, ever larger multiples of prana lead us to units of space and time that allow us to understand the periodic movement of the moon and the planets around the sun. In such a sophisticated multi-layered conception, individual lifetimes of human beings are held to count little and yet to form an important part of kala and aakash.[15]

Lakshana: Distinguishing Attributes

We may best understand the expression of such concepts in the Jantar Mantars by identifying their key lakshana/lakshanam.[16] Literally the signs/marks over different instants of time which help to make perception manifest,[17] lakshana are the attributes that constitute the essential character of something and help us to perceive it. In the context of indigenous design systems, the methodology or vidhi used to create distinguishing physical characteristics deemed essential in different situations is generally passed forward as a structure of technical, philosophical and social practices. The methodology begins with listing lakshana as well as their underlying principles. To retrieve key lakshana of extant works of indigenous design—such as the Jantar Mantars—we thus also need to comprehend their underlying principles.

The foremost lakshana that distinguishes the Jantar Mantars is interconnectedness, at both a conceptual and a practical level. Using the yantras at the observatories requires an observer to scan the skies as a whole; to repeatedly align and orient themselves and their eyes with particular celestial objects; to judge precisely the boundaries of shadows to obtain readings. In doing so, it is impossible not to be constantly aware of the scale of human beings in relation to the universe, and to realize how insignificant we are in comparison to that scale, as well as how vitally we are linked to it. At a more direct level, the siting of the Jantar Mantars is deeply connected with the tradition of Indian astronomy. The areas where they were established are of great religious importance, which in the Indian subcontinent is inextricably linked with astronomy.

At the level of design and use, starting from the markings inscribed on the surfaces of the yantras—subdivisions or multiples of the unit of prana—and extending to their orientation and form, the Jantar Mantars represent defining features of the Indian tradition of astronomy.[18] One such feature is the use of specific *nakshatras* or asterisms as a reference for observations. These nakshatras, the reference groups of stars, appear to move very slowly—a movement caused by the precession or slow backward tilting of the earth's axis. This movement is not observable during ordinary human lifetimes but only in cultures that endure for ages and keep a record of star-positions through these ages, such as the Indian culture. So, ritualistic calendars from Vedic times contain lists of different

The dials and surfaces of Yantras subdivided into time and space markings. Misra Yantra (l), and Ram Yantra (r) at the Delhi Jantar Mantar

nakshatras to mark the equinoxes or the solstices. The periodic revisions in these calendars and their reference nakshatras allow us to calculate with precision the period when they were compiled, and trace the precession back.[19] The other feature is the importance given to the sun, in daily, seasonal, and annual spiritual rites as well as in astronomy. These features are visible in the planning as well as use of the yantra.

The second key lakshana of the Jantar Mantars is their manifold levels of use. In principle and in practice, they represent multiple aspects of ritualistic, speculative, and practical modes of knowledge. The applications of their yantras span religious practice, the cultivation of scientific knowledge, and dissemination of practical information affecting day-to-day lives of people. In the first instance, of course, their function was related to the ritual practices of their chief patron, Sawai Jai Singh, as well as of the rest of the populace. So, the entrance to the Delhi observatory, the first of the Jantar Mantars, was preceded by a temple aligned to the center of its main instruments. Additionally, the observations and their analysis were used to predict forthcoming planetary and star events, eclipses and positions of the new moon, and seasonal weather changes and natural calamities. These in turn were used to plan the agricultural and administrative calendar and determine related economic activity. Such observational and analytical astronomy and knowledge of its climatic, psychological and other effects are in "The Tradition of Indian Astronomy," as described in my book:

> Of the many spheres of knowledge cultivated in the Vedic civilization of India—now increasingly recognized as probably the oldest civilization of the ancient world—astronomy was especially exalted. It was the first and chief of the sciences auxiliary to the sacred scriptures. Among the six *Vedangas*, the limbs of the Vedas, astronomy is represented as the eyes. Such a representation was not merely symbolic. The Vedic calendar, based as it was upon the movement of both the sun and the moon in relation to the constellations, involved detailed observation and analysis of the daytime and night sky. Vedic life was centred on sacrificial rites prescribed for different times of the day and of the year. The correct performance of these rites required Vedic priests to devise precise calendars and construct complex geometrical altars. They were thus, experts in calendaric astronomy and in mathematics and geometry.[20]

The stated purpose for building the Jantar Mantars—the revision of astronomical tables on the basis of which the yearly calendars were made in the Mughal Empire—has to be, therefore, understood as a cultural continuity of the Vedic tradition of sustained co-relation between theoretical calculations and actual positions of celestial bodies to benefit multiple users. The revised set of astronomical tables, the *Zij-i Muhammad Shahi*, named after the reigning Mughal emperor, Muhammad Shah, and ceremonially presented to him by Sawai Jai Singh in 1728 CE, were accordingly used to compute planetary, lunar, and stellar positions and to schedule religious, agricultural, and

Bhairon Mandir aligned to the central Yantras of Delhi Jantar Mantar, surmised to be part of the original entrance sequence at the observatory; a page from a Pancang, the traditional almanac listing auspicious days for ceremonies, festivals, and fasts.

social activities.[21] Many copies were made, and the information in it was also utilized in the Pancangs, the traditional Hindu almanacs. Pancangs are still used to forecast annual weather conditions and to schedule weddings, coming-of-age ceremonies and festivals. Skilled traditional astronomers such as Pandit O. P. Sharma at the Jaipur Jantar Mantar transmit the knowledge of how the yantras may be used for generating precise readings with the unaided human eye, as

well as the applicability of these measurements in astrology.

The third key lakshana of the Jantar Mantar, an immediately perceivable attribute, is that of optimization and decentralization. This lakshana is visible at a tangible level in the formulation of the yantras as multi-purpose instruments, which can be accessed by many users at the same time. For instance, though the Samrat Yantra is essentially a sun-clock, with which skilled observers may record time with an accuracy of two seconds, it is additionally devised to measure the sun's declination. Similarly, the JaiPrakash Yantra is designed to locate the position of the sun, the stars, and the planets during the day and night, through both their local and universal coordinates. It also has a secondary function of determining local solar time.

Not only is one yantra used for more than one function, but also the buildings that compose the yantra often contain more than one instrument. As an example, the high-precision Shasthamsa Yantra designed to measure the parameters of the sun is housed in chambers that are a composite part of the Samrat Yantra of Jaipur

The Yantras of the Jantar Mantar: rare examples of accessible architecture and astronomy that encompass multiple functions

and Delhi. Indeed, the very fact that the yantras of the Jantar Mantars are on the scale of buildings, is because Jai Singh wanted everyone to develop their own powers of observation and analysis. Their design and construction makes the manner of using them relatively simple and accessible to even non-experts with some amount of training and practice—as recent programs with schoolchildren and amateurs at the Delhi Jantar Mantar have shown. This principle is in contrast to instruments such as the telescope—which need to be protected, which can be used by only a limited number of people at a time, and which cause an observer to focus attention on only a part of the sky, and exclude everything else from sight. Jai Singh explicitly states in the *Zij-i Muhammad Shahi*:

> Since the telescope is not readily available to an average man, we are going to base our rules of computation for naked-eye observations only, which, in turn, are based on earlier texts.[22]

Thus, the yantras of the Jantar Mantars, as much works of architecture as they are instruments of astronomy, have braved the elements for three centuries and continue to firmly orient viewers and observers with the solidity of the earth and the expanse of the sky. They combine utility and beauty, theory and practice, and the sacred and the everyday.

Vidhi: Methodology

The identification of the lakshana of the Jantar Mantars—their essential characteristics—lead us on to the *vidhi* or the methodology of making them. This is in line with the classic ways of learning and research in Indian systems, which follow a structure of *purvapaksh, khandana,* and *uttarapaksh/siddhanth*. This roughly translates as: prior/existing view, analysis or refutation, and subsequent stance/conclusive principle. The vidhi is, thus, of an encyclopedic mode. It investigates all sources of information/opinions before formulating

individual responses. An outcome of such a methodology is that there is constant evaluation of existing knowledge and ample space for individual interpretation, rather than mere passive reception or repetition of "facts." The method of devising, making, and using the Jantar Mantars follows such an encyclopedic and democratic methodology.

The Jantar Mantars represent the uttarapaksha or the conclusive principles arrived at, as the finale of wide-ranging research initiated by Sawai Jai Singh. Thorough training in mathematics and astronomy was essential in the Indian culture, and imparted not just to those who specialized in astronomy/astrology but also to princes such as Sawai Jai Singh. He demonstrated an exceptional interest and aptitude for these subjects, and funded, encouraged, and actively sought out scholars from his own tradition of astronomy and from different traditions and faiths to his court. Many books on related subjects were acquired, translated, and composed in his reign. Curious about research in astronomy in other parts of the world, he sent a team to Europe to discover the methods being followed there in the eighteenth century. Thus, Sawai Jai Singh investigated all available systems of astronomy, before formulating the eventual form and size of the yantras in the Jantar Mantars. He personally participated in each stage of the development—Joseph du Bois, a French Jesuit priest and physician, who translated some European astronomy works for Sawai Jai Singh, writes that the Maharaja prepared wax models of instruments with his own hands. Conscientious in acknowledging his influences, and confident of his original contributions, Jai Singh clearly writes of the initial instruments at the Delhi observatory "such as had been constructed at Samarcand agreeably to the Mussalman books" as well as of "instruments of his own invention" in his introduction to the *Zij-i Muhammad Shahi*.

Sawai Jai Singh focused on the establishment of the Jantar Mantars from between the end of the second decade to the third decade of the eighteenth century, at a time when he wielded great influence in the Mughal imperial system and played an active role in the Mughal court at Delhi. Many of the yantras that we see today are original innovations devised after an exhaustive process, even when they are based on principles of earlier instruments. No surviving examples of large masonry instruments existed anywhere in the world to serve as models for the Jantar Mantars. The well-known Samarkand observatory built in 1424 CE by Mirza Ulugh Beg, grandson of the mighty Mongol warrior Timur Leng, and the thirteenth-century observatory at Maragha did not exist in a complete state during the time of Jai Singh. After studying existing small metal instruments in the subcontinent and the principles of large non-extant instruments, different materials, sizes, and forms were experimented with. Many working models were made in wood, stone, or metal before the actual work.

Just as the conception of the Jantar Mantars was the result of extensive teamwork, so was their planning and construction, which was carried out in unison by master-masons, gurus, and astronomers. In a Sanskrit book called the *Samrat Siddhantha*, Pandit Jagannath Samrat, one of the most influential of Jai Singh's advisors and collaborators, explains the way in which these were built. The site was chosen carefully on the basis of certain considerations. It was then prepared appropriately, and the ground beneath each prospective instrument leveled perfectly with the help of channels filled with water—as can still be seen at the Jaipur Jantar Mantar. On the level ground obtained, precise and accurate north-south and east-west directions were laid down by observing the shadow of a vertical rod cast by the sun, as detailed in Hindu astronomical manuals such as the *Surya Siddhantha*.

To make sure that the yantras would be stable and observations of the horizon would not be disturbed, they were constructed partly below the surrounding ground level. Their dimensions were marked

out as full-scale templates on site, and the weight of their huge masonry structures was lessened by scooping out arched niches and openings. They were finished with perfectly aligned and smoothed plaster of lime, whose material and texture is most appropriate for taking shadow-readings, the main method of using the yantras.

Methodology and Construction: A page from a copy of the *Samrat Siddhantha*; a mason at the Jaipur Jantar Mantar demonstrating the special stone used in preparing the final gleaming coat of lime-render; the Jai Prakash Yantra at Jaipur, its dials re-laid in marble (instead of the original lime-render appropriate for shadow-readings)

Their measuring surfaces were finally rendered with multiple layers of fine slow-setting lime, as recorded by Pandit Jagannath and as revealed through plaster investigations and field studies.[23] They were simultaneously calibrated in minute subdivisions of time and subdivisions of angles before the plaster set, requiring complete coordination between skilled astronomers, mathematicians, and masons. It is because of such united and concerted different skills that despite later interventions, earthquakes, and invasions—these huge instruments continue to exist as precise and stable structures, even though many of them have subsequently been re-laid with sandstone and marble or their markings covered over with cement-based paints.[24]

The extraordinary sense of space within and around the Yantras: The Samrat Yantra seen through the Jai Prakash Yantra at Delhi

Discovering the World and Ourselves

While the Jantar Mantars reinforce the characteristics of indigenous knowledge-systems that they are principally derived from, their influence is not limited merely to the Indian subcontinent. They add to the corpus of existing knowledge about astronomy and astrology. As Jai Singh notes in the conclusion to the *Zij*, "with the help of our own new instruments, we have observed five well-known satellites and other heavenly bodies; we have revealed a number of new data."[25] Using these observations, his astronomers were able to discern errors in the *Tabulae Astronomicae* of the French astronomer Phillip de la Hire[26]—the same tables that some writers claim were adopted full-scale by Jai Singh. The yantras of the Jantar Mantars are part of the heritage of our world; they continue to reveal new dimensions and new possibilities, and communicate with an immediacy that hones and directs our perception.

So it is that despite their denuded and changed circumstances through the three centuries of their existence, and despite all the present-day devices of telescopes with micrometer gauges, GPS devices, and artificial satellites, the Jantar Mantars retain a unique power.

Even when we are unfamiliar with the process by which they came into existence or the manner of using them, we are inexplicably moved by a heightened sense of our place in the universe when we are within and around the yantras of the Jantar Mantars. They raise questions in our minds not just about the yantras, but also about the mystery of the world, and indeed the cosmos.

Understanding the context of the profound philosophy of interconnectedness, decentralization, and multiple-use of their world-view helps us to engage more deeply with the Jantar Mantars—and with ourselves; to realize the purpose behind the forms of the yantras, and to look beyond the appeal of their abstract architectural shapes; to comprehend the clarity with which they demonstrate the movement of the earth and the planets, and how they are accessible to an ordinary person to observe the skies and imbibe principles of spherical trigonometry, physics, mathematics, and geography. This is what we need to understand about them; this is why they were made; this is why they need to be experienced and celebrated. They are of immeasurable value to help anyone so inclined to set off on quests of self-discovery and personal interpretation. *Celestial Mirrors* is one such quest, entirely in the spirit of Jai Singh's endeavors.

Notes

1   *The Journal of a Tour in India,* Captain Mundy, 1828, p. 56

2   Penelope Chetwode, "Delhi Observatory, The Paradise of an Early Cubist," *The Architectural Review,* London 1935, p. 57

3   *travel.aol.com/travel...delhi/jantar-mantar-thingstodo-detail-319631/* Cached, accessed 20 Sep 2010

4   V. N. Sharma, Motilal Banarsidass Publishers Private Limited, New Delhi 1995

5   Bharata Manisha Research Series II, First edition Varanasi, 1978

6   A Jyotishacharya is a teacher or expert in *Jyotishigyan,* or Astronomy / Astrology

7   p. 99, V. N. Sharma

8   W. Hunter, *Asiatic Researches*, Vol. Five, "Some Account of the Astronomical Labours of Jayasinha, Rajah of Ambhere or Jayanagar," p. 210

9   The standard reference, written by George R. Kaye, considers these observatories to be "archaeological remains." *The Astronomical Observatories of Jai Singh*, Archaeological Survey of India, New Imperial Series Volume XL, 1918, p. 90

10 The background to most Indian philosophies is the belief that changes of dharma as well as *lakshana* (characteristics) and *avastha* (condition) are in the very nature of the elements that make up the universe. For a brief survey of the meaning and implication of dharma, see https://en.wikipedia.org/wiki/Dharma

11 See A. K. Coomaraswamy, *Spiritual Authority and Temporal Power;* Pujyashri Chandrashekharan Svami, *Hindu Dharma;* Chaturvedi Badrinath, *The Mahabharata, An Enquiry into the Human Condition*

12 p. 827, "Maitri Upanishad," *The Principal Upanishads,* Eng. Translation S. Radhakrishnan, Allen and Unwin, 1953

13 pp. 53 and 310, Phanindralal Gangooly (Ed.), *Surya Siddhanta,* Eng. Translation E. Burgess, 1860, Motilal Banarsidass Publishers Pvt. Ltd., Delhi, 1989, this edition 2005

14 1 prana =10 vipala, and 1 vipala = 60 prativipala. Thus 1 prativipala is equivalent to $1/60$ of a vipala or $1/600$ of a prana. Since 1 prana is equivalent to 4 seconds, a prativipala is $4/600$ or $1/150$th of a second

15 Derived from the root *kas:* that which is visible; aakash (that which is not visible) is also used to indicate the sky

16 *Lakshanam* is the noun derived from the combination of *laksh* (to perceive, to observe) and *kshanam* (instant)

17 *The Concise Sanskrit Dictionary, Sanskrit-Hindi-English,* Compiled by Dr Ram Sagar Tripathi, p. 148

18 As pointed out by the Vedic scholar, David Frawley, in *Gods, Sages, and Kings,* Motilal Banarsidass Publishers Pvt. Ltd., Delhi, 1993

19 In contrast, the "Tropical" zodiac of the older Western world determines the zodiac relative to the Tropics of Cancer and Capricorn, and in fact ignores the precession. Thus, the rate of precession as recorded in the Surya Siddhantha at 54 arc-sec per year is much closer to the current calculation of 50.3 arc-sec per year than that proposed by Hipparchus in the second century BC (36 arc-sec per year). Ptolemy, the great Greek astronomer / astrologer who continued Hipparchus' work on precession, adopted this value.

20 p. 172, *Jantar Mantar, Sawai Jai Singh's Observatory in Delhi,* Ambi Knowledge Resources Pvt. Ltd. 2010

21 A letter, written by a Colonel Pearse in the late eighteenth century, four decades after Sawai Jai Singh's demise, records how he received in Bengal the prediction of three comets and an earthquake three months before the earthquake destroyed extensive regions around Lahore. "On the Sixth Satellite of Saturn," Letter from Col. T. D. Pearse to Secretary, Royal Society London, dated Madras 22 September 1783; p.111 in Dharampal, *Indian Science and Technology in the Eighteenth Century, Collected Writings.*

22 ZMS, Ms. Add. 14373, Dept. of Oriental Manuscripts, The British Library, London, f.189, as quoted in V. N. Sharma, *Sawai Jai Singh and His Astronomy,* p. 243

23 Primary evidence of a lime-pozzolanic mix plaster was found in selective investigations at the Delhi Jantar Mantar. See "Report on the Preliminary Works, Misra Yantra, Jantar Mantar, New Delhi, Part I, Excavations, Plaster and Calibration Investigations, February-June 2007," prepared by Anisha Shekhar Mukherji, July 2007

24 For instance, the Niyat Chakras (on the south face of the Misra Yantra) in the Delhi Jantar Mantar were resurfaced in marble in 1951. The first available reference to the painting of the yantras in terracotta color is found in *Indian Archaeology — A Review (IAR) 1961-2.* See "Misra Yantra: Preliminary Works Part II, Archival Investigations Report September 2008," prepared by Anisha Shekhar Mukherji for The Redevelopment of the Jantar Mantar Observatory Complex, New Delhi

25 G. Sobirov, *Samarkand Scientific School of Ulugh Beg,* Dushanbe, 1975 (in Russian), as quoted in A. Rahman, *Maharaja Sawai Jai Singh II and Indian Renaissance,* p. 39

26 Claude Boudier, the French Jesuit priest who observed the solar eclipse at the Delhi Jantar Mantar with the observatory staff in 1734, was able to note that it did not agree with La Hire's calculations, confirming Jai Singh's earlier assessment; quoted in V. N. Sharma, *Sawai Jai Singh and His Astronomy,* pp. 244 and 300

Explanation

The masonry instruments, which vary in size from a few feet to 90 feet in height, are Jai Singh's chief work.

G. R. Kaye, *The Astronomical Observatories of Jai Singh*

Jai Singh's observatories present a fascinating intersection of astronomy, religion, mathematics, architecture, and design. We begin with a look at the observatory sites and the principles of astronomy that underlie them. We then move on to look at the major instruments, to better understand their design and function. Although the telescope was already in use elsewhere, Jai Singh chose the tradition of naked-eye sky observation as the basis for his observatories, creating architectural forms that are both beautiful and functional.

The Observatories

When Jai Singh built his observatories at the beginning of the eighteenth century, nothing like them had been seen before in India. Whether on the roof of a palace, in a medieval fort, or on open lands, the unique forms of his masonry instruments made a strong impression on visitors and locals alike. Today, in our highly advanced world, these remarkable places still remind us of the enormous mystery that surrounds us in the night sky.

The Delhi and Jaipur observatories are the largest and best known, comprising an area of about 5 acres each. The Jaipur observatory, located just outside the walls of Jai Singh's palace, is the most complete, and includes instruments not found at the other observatories. The Delhi observatory, near the center of the city, has only a few instruments, but they are at very large scale.

Further east, the Varanasi observatory sits on the rooftop level of the Man Singh palace on the banks of the Ganges River, and to the south, the Ujjain observatory is located in the suburbs of that city. The Mathura observatory no longer exists.

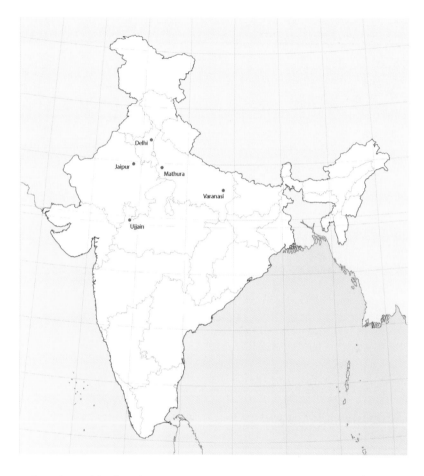

Locations of the five observatories originally constructed by Jai Singh. Some sources suggest that the Maharaja built multiple observatories so that he could compare readings between them, to obtain better accuracy.

Delhi

Jaipur

Ujjain

Varanasi

Site Plans

Site plans provide an overhead view that allows us to see how buildings are laid out, how parts connect to one another and to the larger site on which they are located. Many of the source materials about the observatories include copies of site plans based on measurements taken at the observatories, and modern satellite imagery also reveals the layout of structures as seen from above.

The illustrations in this section combine architectural drawings, outlined in red, with backgrounds taken from satellite imagery to show the precise form and placement of the observatory structures within the larger urban fabric in which they now sit.

The names of the structures that served as astronomical instruments are indicated on each plan, and as you study them, you will recognize the essential forms of the Samrat Yantra, (cross-like) Rama Yantra, and Jai Prakash Yantra (circular forms that come in pairs), and the Digamsa Yantra (concentric circles). Later, we will look at each of these instruments and others in greater detail.

Delhi

Built in 1724 on the outskirts of old Delhi, the observatory was a gift to the Mughal emperor Muhammad Shah. Nested in the heart of downtown New Delhi it also serves as a park and gathering place for political protest.

Jaipur

Situated in the city that Jai Singh designed as the new capital of his kingdom, the observatory at Jaipur is the most complex and elaborate, comprising sixteen different instruments for naked-eye sky observation.

Ujjain

Located in the ancient city of Ujjain, a center for Hindu astronomy since early times, the Ujjain observatory is the southernmost of the five observatories. It is very well maintained and is in active use as an educational center and meterological station for the region.

Varanasi

Built in the city renowned as a center of learning since the time of the Buddha, the observatory is uniquely placed atop the Man Singh palace, along the edge of the Ganges. The Varanasi observatory has by far the smallest footprint, yet incorporates five essential instruments.

Delhi Observatory
ca. 1721-1724

The observatory at Delhi was built in 1724 on lands belonging to Jai Singh just to the south of Shahjahanabad, known as "Old Delhi" today. The observatory was built as a gift to Emperor Muhammad Shah, who controlled much of Northern India, including Jai Singh's native Amber. In the preface to the *Zij-i Muhammad Shahi,* a set of astronomical tables created from readings at the observatory and dedicated to the emperor, Jai Singh sets out his reasons for creating the observatory. He observed that the existing star tables contained significant inaccuracies and gave "widely different results than those determined by observation." The Maharaja went on to explain how he took on the task of producing exact and correct observations through the construction of instruments based on earlier designs such as those at Samarkand.

Located in what is now the heart of modern Delhi, the observatory structures occupy a park-like compound within a few blocks of Connaught Place. When the observatory was originally constructed, the open plains gave an unobstructed view of the sky, but today, tall buildings and trees limit the view of the sky toward the horizon.

Each of the observatories has a unique color scheme. The masonry structures at the Delhi observatory are finished with a red wash and the index surfaces are white plaster, light brown sandstone, or marble, depending on the instrument. The Delhi observatory is maintained by the Archaeological Survey of India.

Masonry Instruments
  Samrat Yantra
  Jai Prakash Yantra
  Rama Yantra
  Misra Yantra
  Shasthamsa Yantra
  Agra Yantra

Location
Latitude:	28° 37' 36.62" N
Longitude:	77° 12' 59.79" E
Local time:	UT + 5 hr, 8 min, 52 sec
Elevation:	725 ft.

Misra Yantra

Samrat Yantra

Jai Prakash Yantra

Rama Yantra

meters
0   10   50   100

Jaipur Observatory
ca. 1728-1738

The Jaipur observatory is by far the most elaborate and complete of Jai Singh's projects, comprising sixteen masonry instruments and six made of metal. The observatory occupies a plot of land just outside the City Palace, within the walls of the original city.

The city of Jaipur was designed and built by Jai Singh in 1727 to become his primary residence, with his palace at the center. He was 39 years old at the time. The Maharaja had been interested in observational astronomy since his early twenties, and the construction of an observatory adjacent to his palace provided a way to test his ideas about large-scale instruments for naked-eye sky observation.

The Jaipur observatory includes a number of instruments that are not duplicated at the other observatories. These include the Kappala Yantra, Rasivalaya Yantras, and Unnatamsha Yantra.

Jai Singh's observatories saw their most active use during the two decades of his reign, and by the early 1730s the Jaipur observatory was the primary site of the Maharaja's astronomical activity. After Jai Singh's death in 1743, the observatory fell into disuse, except for the period 1751-1778 during the reign of Madho Singh. In the late 1700s, a casting facility for making cannons was built within the observatory compound. A major restoration was completed in 1902, and further restorations in the last century replaced many of the plaster scales with marble surfaces.

The Jaipur observatory is managed by the government of Rajasthan and in 2010 it became a World Heritage site. The materials at Jaipur include red sandstone, limestone, plaster, and marble. The plaster is finished with a yellow wash.

Masonry instruments

Krantivrtta
Digamsa Yantra
Rama Yantra
Kapala Yantra
Jai Prakash Yantra
Nadivalaya
Small Samrat Yantra
Palabha
Daksinottara Bhitti
Great Samrat Yantra
Shasthamsa Yantra
Rasivalaya Yantras
Horizontal circular dial
Dhruvadarsaka Pattika
Rama Yantra models
Krantivrtta II

Metal instruments

Cakra Yantra
Yantraraja
Incomplete Yantraraja
Unnatamsha Yantra
Krantivrrta
Samrat Yantra models

Location

Latitude:	26° 55' 29.23" N
Longitude:	75° 49' 29.17" E
Local time:	UT + 5 hr, 3 min, 17 sec
Elevation:	1440 ft.

Unnatamsha Yantra

Yantraraja

Nadivalaya

Daksinottara Bhitti

Small Samrat Yantra

Kapala and Cakra Yantras

Digamsa Yantra

Rama Yantra

Great Samrat Yantra

Jai Prakash Yantra

Rasivalaya Yantras

meters

0   10   50   100

Ujjain Observatory
ca. 1724-1730

Ujjain, an ancient city and center of Hindu astronomy since early times, was also the largest city and capital of Malwa province during Jai Singh's reign. Like the other observatories, it was little used after Jai Singh's death, but underwent a major restoration in 1923. At that time, the markings on the instrument scales were changed from the Hindu system to the Western system of hours, minutes, and seconds.

The observatory is situated on the outskirts of the city in an area that includes residential and agricultural uses. It is well maintained and the instruments are finished in a bright white wash. The observatory also functions as a weather station and educational center for the region.

Masonry instruments

Samrat Yantra

Nadivalaya

Digamsa Yantra

Daksinottara Bhitti

Agra Yantra

Palabha

Sanku

Location

Latitude: 23° 0' 16.59" N

Longitude: 75° 45' 59.55" E

Local time: UT + 5 hr, 32min, 4 sec

Elevation: 1600 ft.

Samrat Yantra

Digamsa Yantra

Nadivalaya

Daksinottara Bhitti

Palabha and Sanku

meters

0   10   50   100

Varanasi Observatory
ca. 1724-1730

The observatory at Varanasi was constructed on the roof of the Man Singh palace, and overlooks the Ganges River. Early accounts of the observatory indicate that it remained in good condition through the late 1700s, but by 1848 its condition had deteriorated considerably. The instruments were restored in 1911, at which time the scale markings were modified to favor Western units of time. While the observatory once had an unobstructed view of the sky, today taller buildings and trees immediately adjacent limit observations close to the horizon. The observatory is managed by the Archaeological Survey of India.

Masonry instruments
Large Samrat Yantra
Digamsa Yantra
Small Samrat Yantra
Nadivalaya
Daksinottara Bhitti I
Daksinottara Bhitti II

Metal instruments
Cakra Yantra

Location
Latitude: 25° 18′ 28.21″ N
Longitude: 83° 00′ 38.88″ E
Local time: UT + 5 hr, 32min, 4 sec
Elevation: 250 ft.

Large Samrat Yantra

Cakra Yantra

Digamsa Yantra

Nadivalaya

Small Samrat Yantra

meters

0   10   50

Looking at the Stars, Lining Things Up

When we gaze upward on a clear night, it's difficult not to be awed by the majesty and mystery of a dark sky filled with countless points of light. For the curious, the night sky is a space rich with patterns and features, planets, stars, and constellations. Although to our eye the pattern seems static, in fact it is always changing—by virtue of earth's rotation on its axis and its orbital movement around the sun.

When we look at the night sky only occasionally, we don't notice the changes. But if we set up references by which to observe the sky— for example a building, or landscape feature—we can make very specific observations about the position and orientation of the things we see in the sky. One evening I may notice that the sun sets at 7:53 p.m. just to the right of a large tree at the end of my driveway. A week later, standing in the same place, I note that it is setting 14 minutes earlier and a few degrees to the left of the tree. The place where I am standing and the tree at the end of my driveway are reference points. That the sun appears to rise and set each day is the result of earth's rotational movement. The apparent change from week to week in the time and place where it sets is the result of earth's orbit around the sun and the tilt of its axis.

Jai Singh used the same observational principle, two reference points, but in much more precise and sophisticated ways. In his solar instruments, he used the light of the sun, traveling in a straight line from sun to earth, to determine the position of the sun in the sky. At night, observers positioned themselves within the structure of his instruments and used sighting guides to align an object in the sky with a point in the instrument that would indicate the object's position.

Crescent moon setting, Ithaca, NY, July 15, 2018, at 9:06 p.m. If I stay in this spot for several minutes I will see the moon's position change relative to the trees and horizon, the result of earth's rotation.

Right: Star trails over Chile's Atacama Desert. The photograph is a time exposure over several hours with the camera pointed toward the sky's south pole. As the earth rotates, the stars appear to move in circular arcs, creating the pattern seen here.
Photo by A. Duro/ESO

Thinking Astronomically

To better understand the instruments and how they were used, it's helpful to know how astronomers conceive of the sky and the positions of the things we see there. If you are familiar with GPS coordinates, or have used a map with a letter-number key, then you know how a coordinate system can be used to identify the location of a place within a larger area.

Astronomers use mainly two different coordinate systems to note the positions of objects in the sky. These are called the *horizontal* or *altitude-azimuth* coordinate system and the *equatorial* coordinate system.

The horizontal coordinate system uses the observer's horizon as a reference and notes the position of a celestial object by its *altitude,* its angular distance above the horizon, and its *azimuth,* the angular distance from north (moving eastward) measured along the horizon. It is easy to use this system with naked-eye sky observation since it is based on the observer's location. It is, however, a *local* measurement; someone else measuring the same sky object from a different location will obtain different coordinates.

The equatorial coordinate system is more abstract, using the center of the earth as a reference and projecting the plane of the equator into space. The plane of the ecliptic, a projection of the plane of the earth's orbit around the sun, intersects the plane of the equator at an angle equal to the earth's tilt on its axis. The points where the two planes intersect represent the equinoxes. The position of an object is noted by measuring eastward around the equatorial plane from the vernal equinox and northward from it. One advantage of this system is that it is independent of the observer's location so the coordinates obtained are universal.

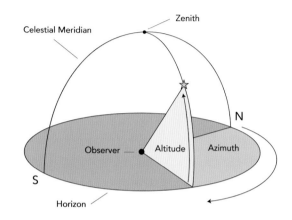

The horizontal coordinate system uses the observer's horizon as a reference and measures a celestial object's position relative to the horizon

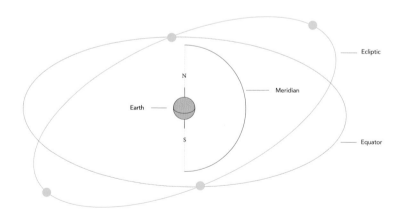

The equatorial coordinate system uses the center of the earth as a reference and measures a celestial object's position based on a projection of the earth's equator into space

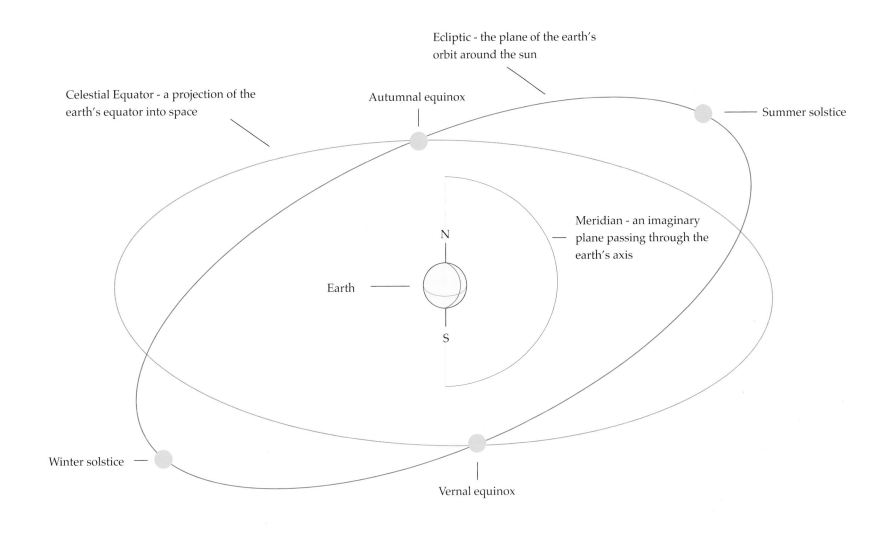

Ecliptic - the plane of the earth's
orbit around the sun

Celestial Equator - a projection of the
earth's equator into space

Autumnal equinox

Summer solstice

Meridian - an imaginary
plane passing through the
earth's axis

N

Earth

S

Winter solstice

Vernal equinox

Diagram of the Celestial Equator, Ecliptic, and Meridian

Solar Observatory
Kapala Yantra

During daytime hours on clear days, the observatory provides accurate measurements of the sun's position. Many of the yantras have measuring functions for both day and night conditions. The Samrat Yantra gives the most precise time in daylight observations, while the Jai Prakash Yantra, which we will consider shortly, indicates other aspects of the sun's apparent movement, including time of day, time of year, solstices and equinoxes.

The Kapala Yantra, pictured at right and on the facing page, is a bowl-shaped instrument like the hemispherical dials of earlier Mediterranean cultures. It was likely a model for the Jai Prakash Yantra.

In this view, the shadow of the sighting guide, suspended by crosswires attached to the rim, can clearly be seen on the marble surface along with inscribed lines that show both horizontal and equatorial coordinate systems. Each system has unique advantages for determining the position of celestial bodies. Additional lines, labeled with devanagari script, map the paths of other celestial objects, including planets, stars, and constellations. Inscribed lines and marks appear on all of the instruments at the observatories. Originally, the instruments at Jai Singh's observatories had inscribed surfaces made of plaster or sandstone. In Jaipur, the earlier plaster surfaces were replaced with marble on many of the instruments, following a renovation in 1945.

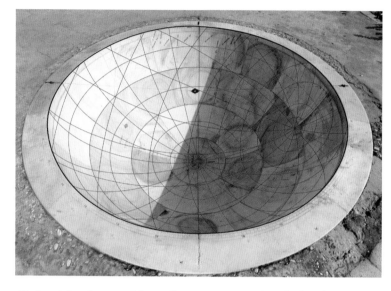

Horizontal and equatorial coordinate systems are insctibed in the surface of the Kapala Yantra. In these illustrations red indicates the horizontal system and blue the equatorial.

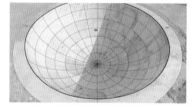 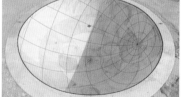

The horizontal coordinate system, above left, and equatorial coordinate system, above right. In the horizontal system, the center point, at the bottom of the hemisphere, corresponds to a point in the sky directly overhead. In the equatorial system, the center point, 2/3 of the way up toward the rim at the right, is a projection of the north pole and aligns with Polaris, the pole star. In both systems, the radial lines represent equal units of time.

Right: Kapala Yantra at the Jaipur observatory

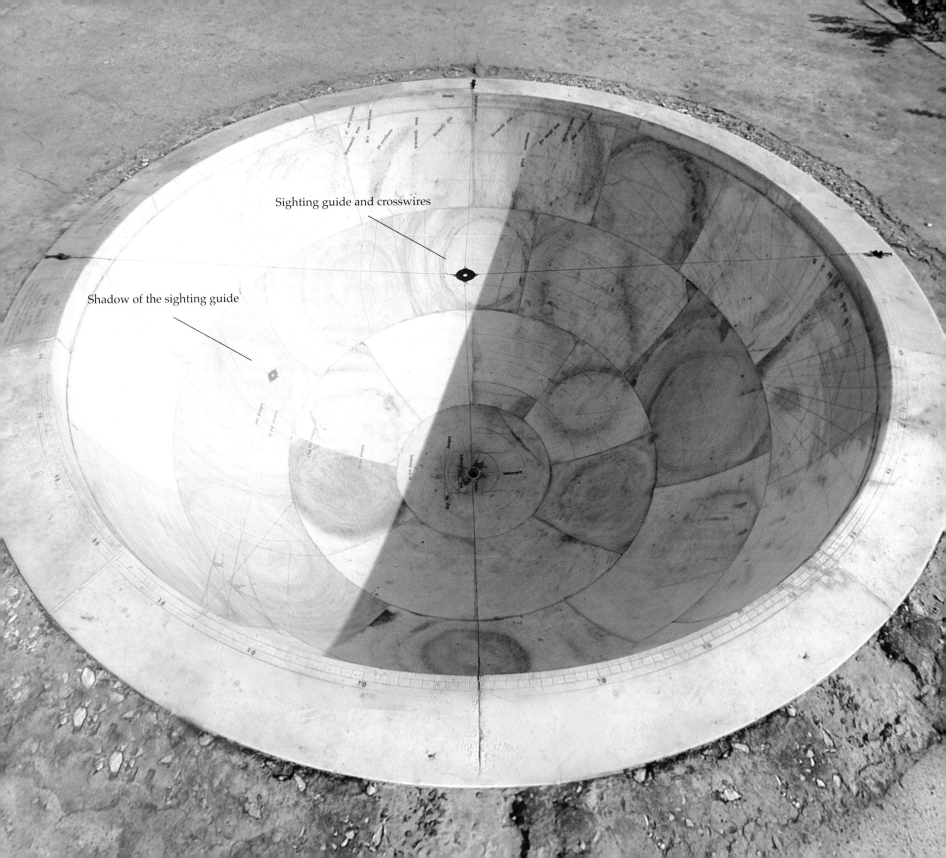

Sighting guide and crosswires

Shadow of the sighting guide

Day/Night Observation
Jai Prakash Yantra

The Jai Prakash Yantra, one of Jai Singh's most beautiful and complex instruments, serves as an excellent illustration of how both day and night sky observation is done. Built only at the Jaipur and Delhi observatories, it consists of a pair of large hemispheres. As in the Kapala Yantra, wires are stretched across the top of each, with a sighting guide at their crossing. The sighting guide thus sits directly over the center of the hemisphere.

The surface of each hemisphere is divided into segments corresponding to 15° of arc, or one hour of time. There is a 15° gap between segments to allow an observer to walk between them, and the two structures complement one another—where one has a scaled surface, the other has a gap. As the observer follows a sky object over time, its corresponding position on the instrument will reach the edge of the scale. When it does, the observer moves to the other instrument, where the object's position will now appear at the edge of the next segment. For solar observation, the astronomer notes the location of the shadow of the sighting guide (visible toward the center of the image on the right-hand page). As in the Kapala Yantra, the surfaces are inscribed with both horizontal and equatorial coordinate systems (visible as fine grid lines in the photograph). Additionally, a number of special or recurring paths, such as the sun's declination at the solstice, and the trajectories of certain planets and constellations, are inscribed and noted in devanagari script.

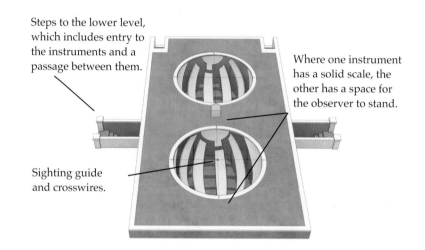

Steps to the lower level, which includes entry to the instruments and a passage between them.

Where one instrument has a solid scale, the other has a space for the observer to stand.

Sighting guide and crosswires.

The design of the Jai Prakash Yantra as a complementary pair can easily be seen in these illustrations. Carefully compare the location of the marble scales in each instrument. As the tracking of a celestial object progressed and it reached the edge of the scale, the observer would walk to the other instrument to continue tracking.

Shadow of the sighting guide.

Observing the Night Sky

Mars

Crosswires

In this illustration an astronomer stands inside the half-sphere bowl of the Jai Prakash Yantra at the Jaipur observatory, holding a simple alignment device. Above him, suspended from crosswires attached to the rim of the Yantra, is a metal plate with a circular opening in its center. The plate is suspended directly over the center of the bowl. The astronomer adjusts his position so that a celestial object, in this case Mars, appears inside the circle of the metal plate. This establishes the line-of-sight for Mars. He then picks up the sighting guide and, aligning it so that the celestial object is still visible through the hole in the aperture plate, brings the end into contact with the marble surface of the Yantra. This marks the exact location of the celestial object. The surface of the Jai Prakash Yantra is inscribed with both horizontal and equatorial coordinates, allowing the astronomer to obtain a precise location for the object in either system at the moment of observation.

Determining the exact locations of stars and planets at specific times was crucial to Jai Singh's aim of generating new and more accurate tables for astrologers. The extremely large scale of instruments such as the Jai Prakash Yantra, Rama Yantra, and Samrat Yantra afforded a precision not possible with brass instruments such as the astrolabe and sextant.

The Instruments

When Jai Singh designed the observatories, one of his foremost objectives was to create astronomical instruments that would be more accurate and permanent than the brass instruments in use at the time. His solution was to make them large, sometimes really large, and to make them of stone.

Jai Singh built 15 different types of instruments. Many were based on earlier Hindu and Arabic designs, but seven were of his own invention. In all, he constructed 39 masonry instruments.

The Samrat Yantra is probably one of the best-known instruments because of its enormous size. It is an equinoctial sundial, and the largest ones at Jaipur and Delhi can have an accuracy within 2 to 3 seconds.

An instrument comprising two hemispherical bowls, each with a crosswire and sighting plate suspended across the top, for making observations from below. The Jai Prakash Yantra is an example of the "complementary pairs" unique to Jai Singh and was built only at the Jaipur and Delhi sites.

Another of Jai Singh's remarkable "complementary pairs," the Rama Yantras are cylindrical structures for measuring altitude and azimuth of celestial bodies. They also were built only at Jaipur and Delhi.

Samrat Yantra

The Samrat Yantra or "king of instruments" is an equinoctial sundial, meaning that its scale shows units of time equally spaced. It uses a triangular wall called a *gnomon* to cast a shadow on a curved surface called a *quadrant*. There are two quadrants, one on each side of the gnomon. They represent a quarter-circle arc, and are supported by a masonry structure that angles them to be perpendicular to the earth's axis. The illustrations on this page show how the Samrat Yantra works.

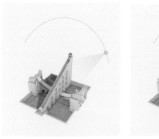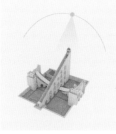

As the sun moves overhead, the shadow of the gnomon moves in the opposite direction along the quadrants. The illustrations above show (l. to r.) 7 a.m., noon, and 2 p.m.

The shadow movement begins at the top of the west (left) quadrant at sunrise, moves to the bottom by noon, and reaches the top of the east (right) quadrant at sunset.

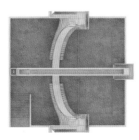

Plan and elevation views of the Samrat Yantra at the Jaipur observatory

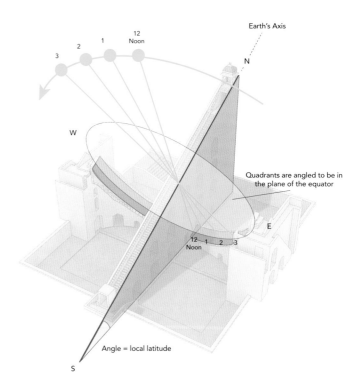

The diagram above shows the *equinoctial* design of the Samrat Yantra. The gnomon, a triangular wall shaded blue, is aligned north-south. Its edge, dark red, is parallel to the earth's axis. The *dial* or time scale, shaded light red, is in the form of two quadrants, or quarter-circles, angled to be perpendicular to the earth's axis and parallel to the plane of the equator. With this design, units of time are equally spaced along the dial. The elevation angle of the gnomon always corresponds to the latitude of the sundial's location.

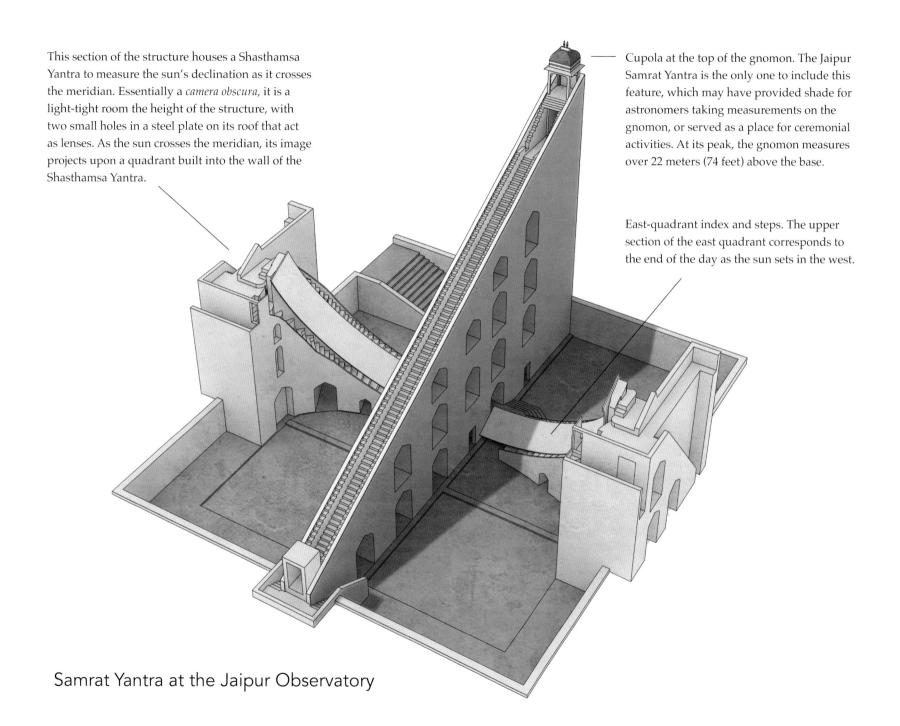

This section of the structure houses a Shasthamsa Yantra to measure the sun's declination as it crosses the meridian. Essentially a *camera obscura*, it is a light-tight room the height of the structure, with two small holes in a steel plate on its roof that act as lenses. As the sun crosses the meridian, its image projects upon a quadrant built into the wall of the Shasthamsa Yantra.

Cupola at the top of the gnomon. The Jaipur Samrat Yantra is the only one to include this feature, which may have provided shade for astronomers taking measurements on the gnomon, or served as a place for ceremonial activities. At its peak, the gnomon measures over 22 meters (74 feet) above the base.

East-quadrant index and steps. The upper section of the east quadrant corresponds to the end of the day as the sun sets in the west.

Samrat Yantra at the Jaipur Observatory

Telling Time
Samrat Yantra

With today's technology it requires little effort to know what time it is. But in the early 1700s, mechanical clocks were not widely available, and needed another clock or timepiece in order to be reset when they wound down or lost accuracy. Sundials and astrolabes were more common timekeeping devices, and always relatively accurate since they relied only on the earth's rotational movement.

Sundials can take many different forms, though the most common are the horizontal sundials often seen as a decorative element in gardens, and the vertical sundials incorporated in the facades of churches and public buildings throughout Europe. All have in common a fixed element such as a rod or triangular plate (*gnomon*), aligned with the earth's axis, that casts a shadow on a surface marked with divisions for the hours and minutes.

When the sun is out on a clear day, the gnomon casts a shadow on the dial, and the shadow indicates the exact time given as *local time*. The term *local time* refers to the fact that since the late nineteenth century, time has been adjusted and standardized by time zones. The earth, however, is in continuous rotation, and events such as sunrise and sunset will occur at different times at different locations within any time zone. Sunrise in New York City occurs about a half hour earlier than it does in Cleveland, and the same difference occurs between Varanasi and Jaipur. The differential in local time is four minutes for each degree of longitude difference between any two places.

Noon has traditionally been used as a reference for timekeeping, so if there were a sundial in each city, they would show noon (the moment when the sun passes directly overhead, and the gnomon casts no shadow) at different times. This would be noon local time, and it would be most correct, but a system based on local time would be impossibly problematic for activities other than local ones.

Samrat Yantras were built at each of the observatories, and both the Jaipur and Varanasi observatories feature two versions of different sizes. The largest, at Jaipur and Delhi, indicate time to an accuracy of about 2 seconds.

Sundials usually have a fixed rod or triangular plate (gnomon), aligned with the earth's axis, that casts a shadow on a surface marked with divisions for hours and minutes. Most versions, such as those pictured above mark only the hours. Although the shadow of the gnomon is sharp on a clear sunny day, it is only possible to estimate the minutes between the hours since they are not indicated on the scales. Photo above left by Anna Armbrust/Pixabay. Photo above right by Mark Caldicott/Pixabay

Right: The western quadrant of the Samrat Yantra at the Jaipur observatory. The marble surface is inscribed with multiple time scales, with the finest scale indicating intervals of 2 seconds.

West quadrant of the Samrat Yantra at the Jaipur observatory. The shadow indicates a time in the late morning, approximately 11:30 a.m.

Shasthamsa Yantra

Built within the towers that support the quadrant scales, the Shasthamsa Yantra gives extremely accurate measurements of the position and size of the sun. It is essentially a *camera obscura* with two aperture plates on the roof of the chamber. At noon local time, as the sun passes overhead, its image is projected by each of the apertures onto quadrants built into the walls of the chamber. Markings on the quadrants enable an observer to measure the sun's declination, or angle above the horizon, and its size.

Image of the sun projected on the quadrant of the Shasthamsa Yantra at the Jaipur observatory

Shasthamsa Yantra at the Jaipur observatory. The sun is just beginning to cross the meridian and project its image on the quadrants.

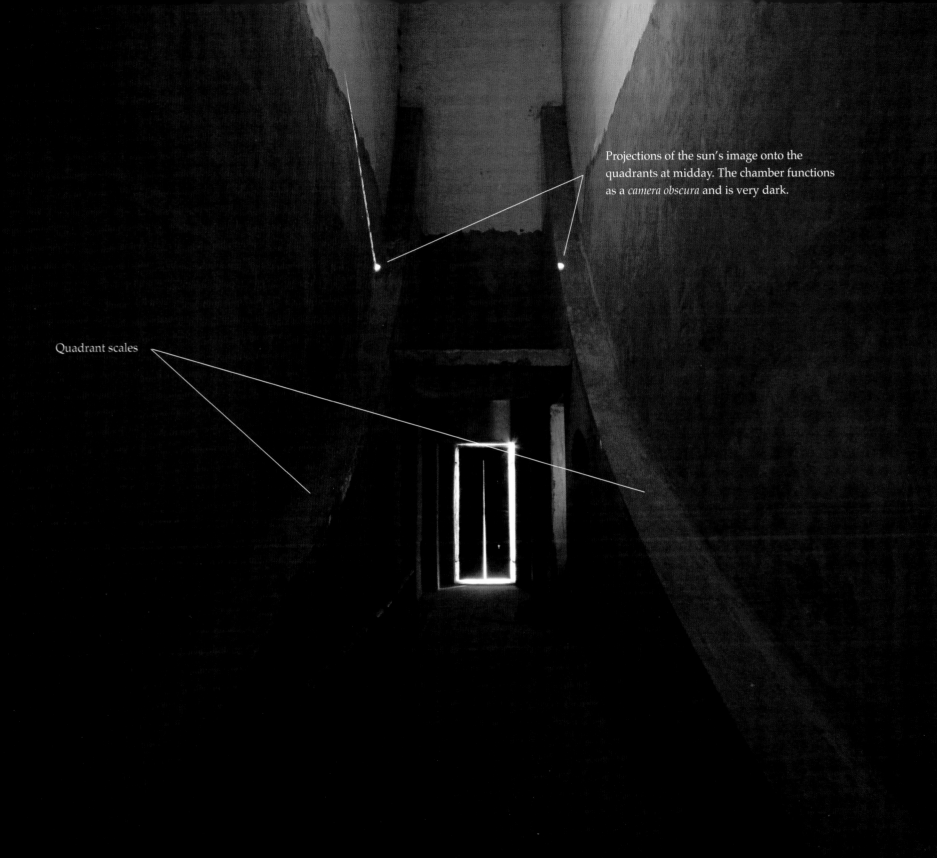

Projections of the sun's image onto the
quadrants at midday. The chamber functions
as a *camera obscura* and is very dark.

Quadrant scales

Jai Prakash Yantra

The Jai Prakash Yantra may well be Jai Singh's most elaborate and complex instrument. It is based on concepts dating to as early as 300 BC when the Greco-Babylonian astronomer Berosus is said to have made a hemispherical sundial. Hemispherical dials also appear in European church architecture during the Middle Ages, and at the observatory in Nanking, China, in the late thirteenth century. The Jai Prakash, however, is much more elaborate, complex, and versatile than any of its predecessors.

The Jai Prakash is a bowl-shaped instrument, built partly above and partly below ground level, as can be seen in the drawings at right. The diameter at the rim of the bowl is 17.5 feet for the Jaipur instrument, and 27 feet at Delhi. The interior surface is divided into segments, and recessed steps between the segments provide access for observers. A taut crosswire, suspended at the level of the rim, holds a metal plate with circular opening directly over the center of the bowl. This plate serves as a sighting device for night observations, and casts an easily identifiable shadow on the interior surface of the bowl for daytime observation of the sun's position. The surfaces of the Jai Prakash are engraved with markings corresponding to an inverted view of both the horizon and equatorial coordinate systems used by astronomers to describe the position of celestial objects.

Among Jai Singh's many contributions to sky observation, perhaps the greatest was the design of paired instruments such as the Jai Prakash Yantra and Rama Yantra at the Jaipur and Delhi observatories. These instruments incorporate inscribed surfaces with a space between them for an observer to stand to take readings. The instruments were exact complements of one another—where one had an inscribed surface, the other would have an empty space

for an observer to stand. If you could lift one and superimpose it over the other, the surface would be continuous, since where one had a void, the other would have a solid surface. In the Jai Prakash Yantra, the hemisphere is divided into sectors representing one hour of observation (15 degrees). As the object being observed reaches the edge of one sector, the observer simply has to walk to the other instrument to continue the observation. See more about this, including time-lapse videos, at *www.jantarmantar.org*.

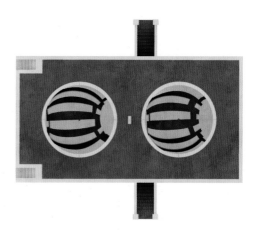

Plan view. The idea of paired instruments for continuous observation is unique to Jai Singh.

Section view of the Jai Prakash Yantra at the Jaipur observatory

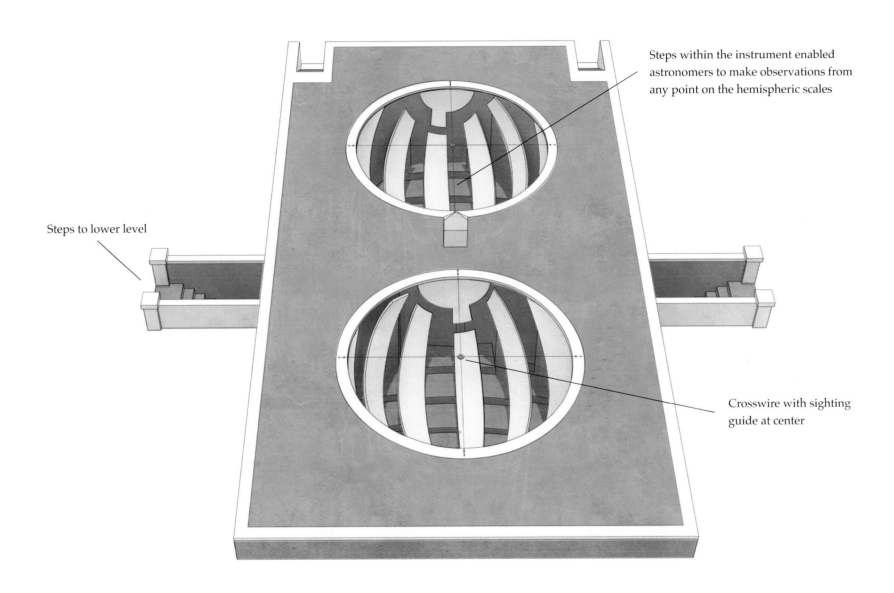

Steps within the instrument enabled
astronomers to make observations from
any point on the hemispheric scales

Steps to lower level

Crosswire with sighting
guide at center

Jai Prakash Yantra at the Jaipur Observatory

Hourly Walk

In the lower level of the Jai Prakash Yantra at Jaipur, we see the passageway that leads from the south instrument to the north instrument. This passageway is fundamental to the design of the Jai Prakash Yantra. As a "complementary pair," the indexed surfaces of one instrument correspond to voids in the other instrument and vice versa. Each of the indexed surfaces represents 15 degrees of the earth's rotation, or one hour of time.

In the course of making observations, astronomers would follow this passageway from one instrument to the other, hour by hour. When the celestial object they were tracking reached the edge of the scale in the instrument they were in, they would simply walk to the other instrument to continue their observation.

Path to the north Jai Prakash Yantra at the Jaipur observatory. To the right at the end of the passageway there is an opening into the bowl of the north instrument.

Rama Yantra

The Rama Yantra consists of a pair of cylindrical structures open to the sky, each with a pillar at the center. The pillar and walls are of equal height, which is also equal to the radius of the structure. The floor and interior surface of the walls are inscribed with scales indicating angles of altitude and azimuth as used in the horizontal coordinate system. Rama Yantras were constructed at the Jaipur and Delhi observatories only.

The Rama Yantra is used to observe the position of any celestial object by aligning an object in the sky with both the top of the central pillar and the point on the floor or wall that completes the alignment. In the daytime, the sun's position is directly observed at the point where the shadow of the top of the pillar falls on the floor or wall. At night, an observer aligns a star or planet with the top of the pillar and uses a sighting guide to determine the point on floor or wall that completes the alignment.

The floor is constructed as a raised platform at chest height, and is arranged in multiple sectors with open spaces between them. This provides a space for the observer to move about and comfortably sight upward from the inscribed surface.

Like the Jai Prakash Yantra, the Rama Yantra was designed as a complementary pair. The Rama Yantras at Delhi and Jaipur differ both in size and in design. The Delhi instruments have an inside diameter of approximately 54 feet and height of 25 feet and incorporate 30 radial sectors, each 6 degrees wide and separated by an equal amount. The Jaipur instruments have an inside diameter of 23 feet and are 15 feet high. They each incorporate 12 radial sectors, but they are not equal in width. Radial sectors in the instrument to the east are 12 degrees wide and separated by 18 degree openings.

The sectors on the instrument to the west are 18 degrees wide with 12 degree spaces between them. The instruments still function as a complementary pair however, and if superimposed the sectors would form one continuous plane.

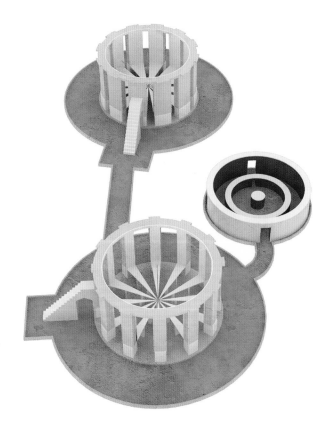

The Rama Yantra and Digamsa Yantra at the Jaipur observatory. Note the difference in width of the radial sectors for the two Rama Yantra units.

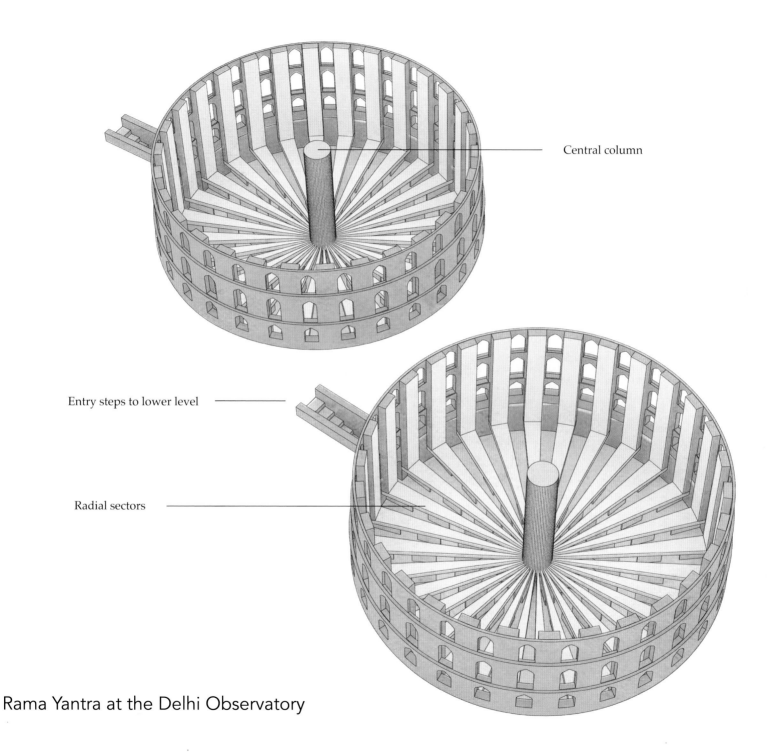

Central column

Entry steps to lower level

Radial sectors

Rama Yantra at the Delhi Observatory

Unlike the Samrat Yantras, which use a curved surface in the equatorial plane to mark the position of celestial objects, the Rama Yantra uses horizontal and vertical surfaces to index visual alignments. This means that the scales must be graduated to compensate for the angular change as the object sighted moves from horizon to zenith. The widest divisions and greatest accuracy occur at 45 degrees, and the divisions become smaller and accuracy diminishes as the object being observed approaches the zenith or horizon. This can be easily seen in the markings on the wall sections in the photograph at right.

To make observations of objects close to the horizon, astronomers had to reach the upper part of the wall. Slots were built into the supporting columns to accommodate a strong shelf or platform for the astronomer to perch on while taking readings.

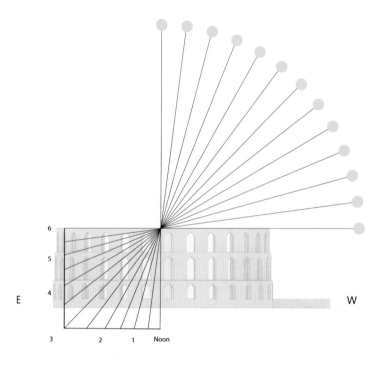

Diagram of the afternoon progression of the sun in 1/2-hour intervals. Note that the divisions grow larger near 3 p.m. (45 degrees), giving the instrument slightly more accuracy at mid-morning and mid-afternoon.

Right: Rama Yantra at the Jaipur observatory

Rasivalaya Yantra

The Rasivalaya Yantras are unique to the Jaipur observatory, and probably the most intriguing of all of Jai Singh's instruments. There are twelve instruments in all, grouped in front of the southern wall of the observatory. The design of the Rasivalaya Yantra is based on the Samrat Yantra, an equinoctial dial with gnomon aligned to the earth's axis and quadrants parallel to the equator. This can easily be seen in the basic form of each instrument. What is also obvious is that although they share this common form, each instrument is shaped and oriented differently from the others. This feature is explained by the instruments' function.

Each Rasivalaya Yantra is related to one of the constellations of the zodiac; thus there are twelve of them, and mounted on each instrument is a painting with a representation of the constellation it is based on.

The gnomon and quadrants of each instrument are aligned with the pole and plane of the ecliptic at the moment the first point in the constellation crosses the meridian. The angles of the gnomons vary from 3.5 degrees (nearly flat) to 50.5 degrees (quite steep), and their orientation varies from north by 26 degrees in each direction.

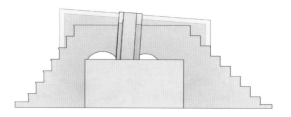

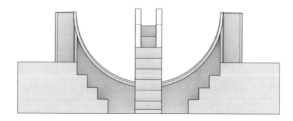

Elevations of the Rasivalaya Yantra oriented to the constellation Cancer. Note that the gnomon is nearly horizontal, pointing toward the pole of the ecliptic when the first point in the constellation crosses the meridian.

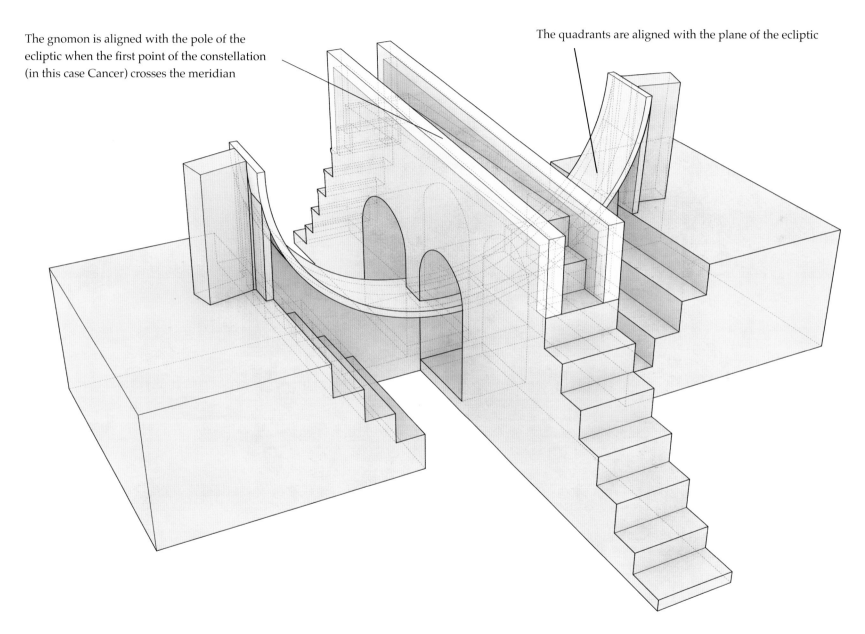

The gnomon is aligned with the pole of the ecliptic when the first point of the constellation (in this case Cancer) crosses the meridian

The quadrants are aligned with the plane of the ecliptic

Rasivalaya Yantra Corresponding to the Constellation Cancer at the Jaipur Observatory

41

Instrument Designer

The Rasivalaya Yantra gives evidence of Jai Singh's deep interest in both astronomy and astrology. It also makes clear that he was most creative and inventive when it came to designing instruments for naked-eye sky observation. His focus may not have been on the mysteries of the night sky, as other astronomers have pondered over the ages, but his passionate interest in seeing more precisely where things were has given us an extraordinary and lasting monument to our place in the cosmos.

Depiction of the constellation Leo mounted on the corresponding Rasivalaya Yantra at the Jaipur observatory

Four of the twelve Rasivalaya Yantras at the Jaipur observatory

Digamsa Yantra

The Digamsa Yantra consists of two concentric cylindrical walls surrounding a central pillar. The top surfaces of the inner and outer walls are marked in angular divisions as fine as 6 minutes of arc.[1] The instrument is used to determine the azimuth of a celestial object by sighting the object from the outer edge of the inner wall through the center of a crosswire suspended from the outer wall, and adjusting a weighted string attached to the central pillar and suspended over the outer wall until it also lines up with the sighted object and center of the crosswire. The weighted string indicates a vertical plane through the intersection point, and the azimuth may be read by looking at the point on the outer wall where the string goes over the edge.

Jai Singh built Digamsa Yantras at Jaipur, Ujjain, and Varanasi, and although they are approximately the same in overall size, the height of the inner wall varies significantly. At Jaipur and Varanasi, the inner wall and pillar are about a meter in height, and the outer wall is twice that. At Ujjain the inner wall and outer wall are the same height. Although the original design called for the outer wall to be twice the height of the inner wall, in practice it only matters that the outer wall be higher than the inner wall.

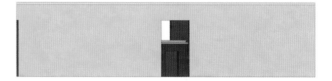

Elevation and section of the Digamsa Yantra at the Jaipur observatory.

1. Standard divisions of a circle are 360 degrees, with each degree divided into 60 minutes and each minute divided into 60 seconds. An arc is a segment of a circle normally described by number of degrees, minutes, seconds.

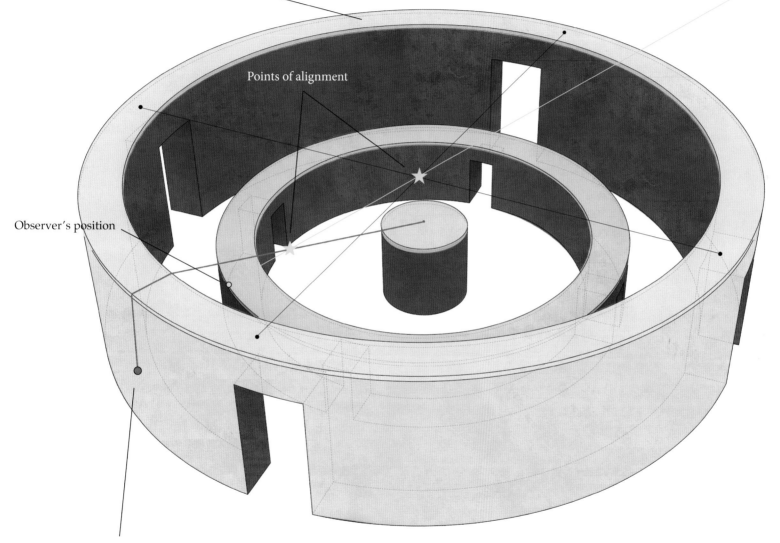

The top of the outer wall is marked in angular divisions of 6 minutes of arc

The observer, standing at the outer edge of the inner wall, first aligns the star with the intersection of the crosswires to create a line of sight. An assistant then moves the weighted string until it intersects the line of sight. The azimuth of the object is then read directly from the point along the top of the outer wall where the string crosses over.

Points of alignment

Observer's position

Weighted string attached to central column and suspended over outer wall is moved to intersect the line of sight already established

Digamsa Yantra at the Jaipur Observatory

Concentric Cylinders

One of the strong impressions from a visit to the observatory is the recurring appearance of elemental geometric forms. The cylinder is so much a part of our lives, in both natural and manmade forms, that we may overlook its simple elegance. When applied to architecture as a way to define and contain space, it can be a poweful symbol of centering, and when the forms are concentric, as in the Digamsa Yantra, ideas of relationship and levels are evoked. Walking the inside perimeter of the outer wall, or stepping close to the edge of the inner wall to peer into the center, brings these ideas into physical experience.

Digamsa Yantra at the Jaipur observatory.

Daksinottara Bhitti

The Daksinottara Bhitti is based on the meridian[1] dial of earlier times. It incorporates a quadrant or half-circle inscribed on a north-south wall, with a metal rod at the center. At midday, the sun casts a shadow of the rod onto the quadrant scale, giving the meridian altitude. The instrument may also be used at night to obtain the meridian altitude of other celestial bodies. Although seemingly basic, meridian-altitude measurements can be used to determine other astronomical data such as local latitude and the obliquity of the ecliptic.

Jai Singh must have considered the Daksinottara Bhitti an important instrument as he built it at all of his observatories. Those at Varanasi and Ujjain include two arcs, and the instrument at Jaipur has three.

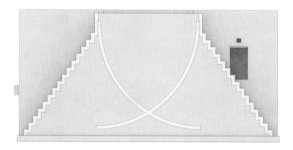

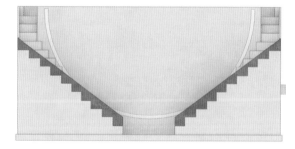

East and west facades of the Daksinottara Bhitti at Jaipur. The instrument incorporates a semicircular arc on the west facade and a pair of overlapping quadrants on the east facade.

1. The meridian is an imaginary plane running north-south in line with the earth's axis and extending directly overhead.

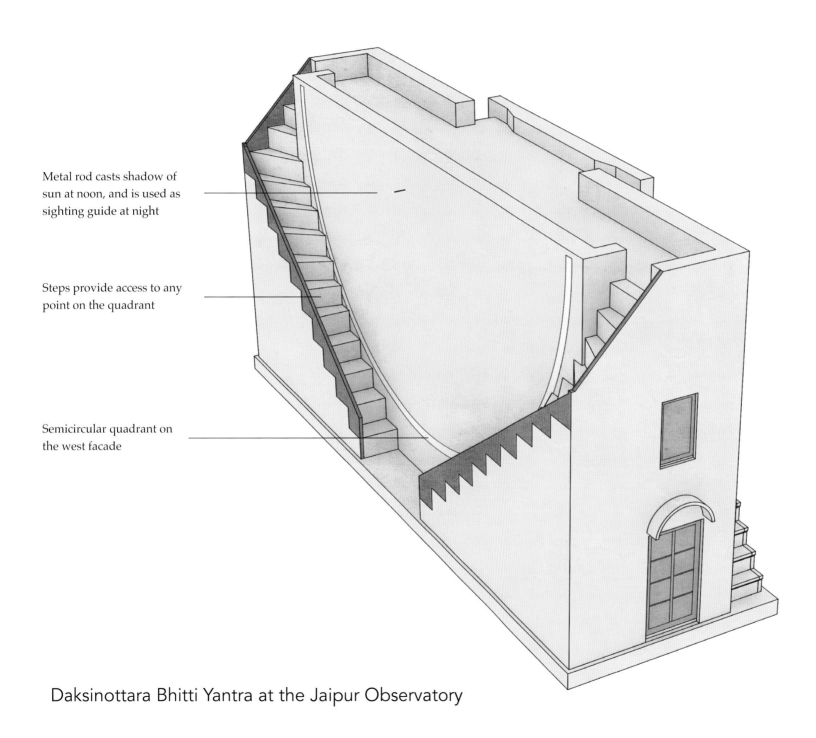

Metal rod casts shadow of
sun at noon, and is used as
sighting guide at night

Steps provide access to any
point on the quadrant

Semicircular quadrant on
the west facade

Daksinottara Bhitti Yantra at the Jaipur Observatory

Nadivalaya

The concept for the Nadivalaya was already known to the astronomers of Jai Singh's time, and references appear in Hindu literature as
early as the 8th century. It is essentially an equinoctial sundial built
in two parts—one facing south and the other north. The dial faces
are built along an east-west line, and are parallel to the plane of the
equator. With this arrangement, the sun illuminates the northern face
during the summer months and the southern face during the winter
months. When the sun crosses the celestial equator, its rays will be
parallel to the planes of the Nadivalaya and will illuminate both faces.
On all other days only one or the other face will be illuminated. This
feature makes the Nadivalaya an excellent indicator of the equinoxes.
The surfaces of the instrument are graduated along the outer edge in
units of time, with the smallest division being one minute. Zero is
indicated at top and bottom, so that time can be measured from noon
or midnight.

The Nadivalaya may also be used for determining the declination
of objects at night. A string is attached to the central pin and extends
to the scale at the rim of the instrument. The string is held taut and
a metal strip or slit is aligned with the object being observed, the
central pin, and the string. The point on the graduated scale where
the string crosses will indicate the hour angle of the object.

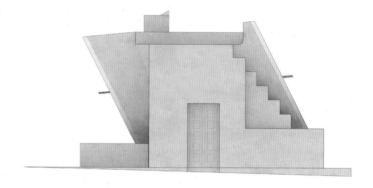

East and south elevations of the Nadivalaya at the Jaipur observatory

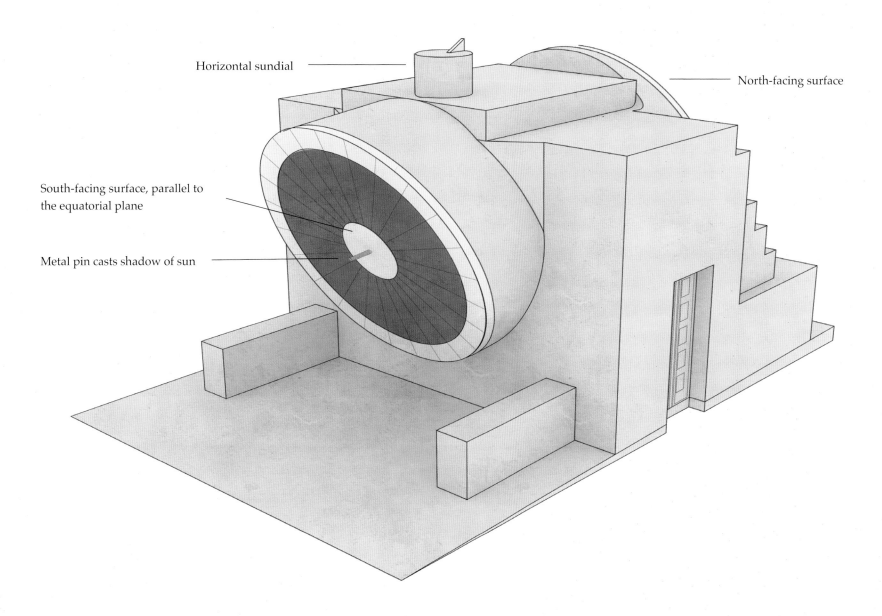

Horizontal sundial

North-facing surface

South-facing surface, parallel to
the equatorial plane

Metal pin casts shadow of sun

Nadivalaya at the Jaipur Observatory

Circles

The circle, another elemental form, appears in instruments like the meridian dials and quandrants of antiquity. Even these simple instruments take on a kind of power when built large. The southern face of the Nadivalaya at Jaipur, roughly twelve feet in diameter and canted forward at about 26 degrees, seems to loom overhead as you approach it.

South face of the Nadivalaya at the Jaipur observatory

Misra Yantra

The Misra Yantra is another unique instrument and was built only at the Delhi observatory. It was neither designed nor built by Jai Singh but is included here because it is a significant part of the overall impression of the Delhi observatory and is so complex. It is also one of the most photographed of the instruments, no doubt because of its beautiful heart-shaped form.

The Misra Yantra was built by Jai Singh's son Madho Singh in the early 1750s and is in fact a compound instrument comprising a Daksinottara Bhitti, Karka Rasivalaya, Laghu Samrat Yantra, Niyat Chakras, and an inclined quadrant.

Elevation and plan views of the Misra Yantra

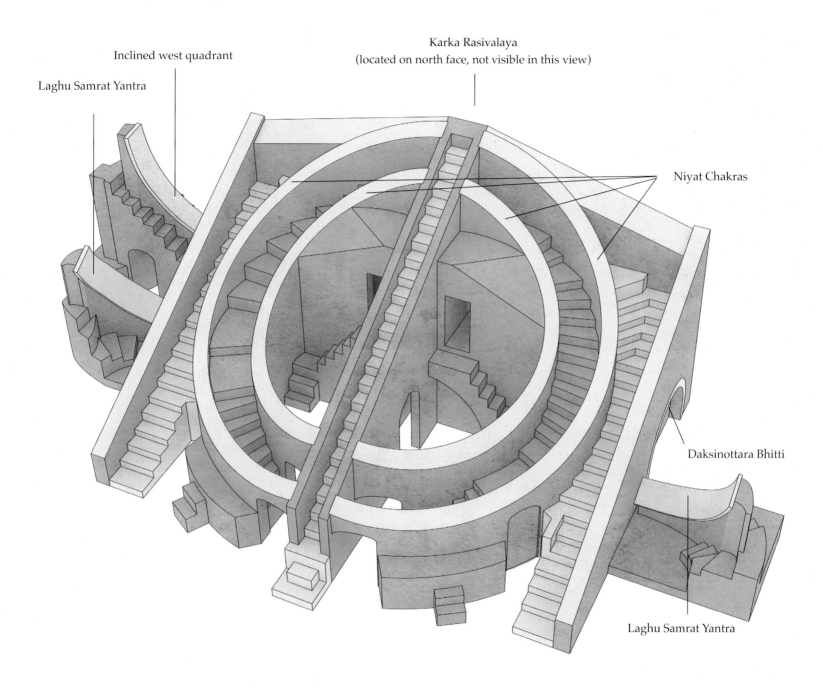

Laghu Samrat Yantra

Inclined west quadrant

Karka Rasivalaya
(located on north face, not visible in this view)

Niyat Chakras

Daksinottara Bhitti

Laghu Samrat Yantra

Misra Yantra at the Delhi Observatory

Immersion

Jai Singh's observatories are unlike any other place in the world. The instruments, with their graceful yet precise geometries, give the impression of an architecture both modern and timeless. Conventional photographs convey what they look like, but don't compare to actually being there.

This section takes you inside Jai Singh's observatories via spherical panoramas originally made for virtual tours at *www.jantarmantar.org*. Capturing a view 360 degrees in every direction, they provide a way to see the observatories in a new way—almost as if you were there.

Spherical Panoramas

The photographs in this section are made from spherical panoramas, so called because they present the scene they capture as though it was displayed on the inside of a sphere. Imagine looking all around you and taking pictures of everything you see—in front, behind, above, below, left, and right. Then imagine you are standing inside of a large sphere with scotch tape and big prints from all these photos. You tape them to the inside of the sphere so that they fit together and overlap to reconstruct the original scene. This assemblage of photos on the inside of a sphere would be a spherical panorama.

This is essentially how VR panoramas are made, and when we look at them on computers, mobile devices or VR viewers, we see a small portion of the sphere that simulates what we would see naturally. As we swipe or drag to navigate, or turn our head with a VR viewer, different areas of the sphere come into view, giving the impression that we are "looking around."

In this book, I take a different approach and present the full spherical panorama as a rectangular image, but in doing so I can't have a perspective that looks correct. That's because when I take the information from the surface of a sphere and present it as a flat rectangle, it has to be *mapped* to fit the new format. The problem is well known to cartographers, and I suspect you will recognize it from the diagrams that follow.

The projection, or mapping, I use is called *equirectangular,* which means that the tapered longitudinal (vertical) sectors on the sphere are projected as equally spaced vertical rectangles. Since the longitudinal sectors narrow to a point at top and bottom, they have to be stretched in the projection to maintain equal width toward top and bottom. This introduces a distortion that we are not used to seeing, though you've probably seen it in world maps using equirectangular projection.

The spherical panoramas in this section take you into Jai Singh's

observatories in an immersive way, but without VR technology. While the odd shapes and curvature may seem alien at first, every scene includes multiple views that look just as they would through an ordinary camera. The photo on the facing page serves as an example, and on the next pages you will see how multiple views may be contained within a larger image. With a little bit of practice you will be able to pick them out for yourself, and with a little more time, you may very well be able to immerse yourself in the space the photo depicts!

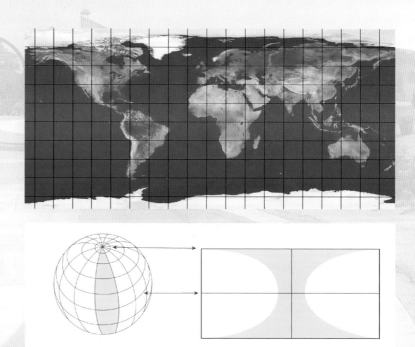

Top: A world map based on equirectangular projection. Land masses toward the north and south poles become increasingly enlarged.

Bottom: The diagram illustrates how vertical sectors are stretched as they are mapped to conform to the 1:2 rectangular format.

This is a view of the Jaipur observatory, taken from the Kapala Yantra, a pair of marble-lined hemispheres sunken into a stone plaza. The view above is looking to the east and includes (l to r) the small Samrat Yantra, Nadivalaya, Cakra Yantra, Kapala Yantras, Great Samrat Yantra with its triangular gnomon and cupola in the distance, and the array of small structures comprising the Rasivalaya Yantras just in front of it and to the right. This view is very wide, and encompasses a field of view of about 120 degrees.

The photo above is from the same location, taken at the same exact moment as part of a spherical panorama. This view is looking towards the west. The sun is visible low in the sky. Long shadows are cast from the Rama Yantra as a family pauses during their visit to the observatory. Visible in this scene are the two Rama Yantras and the Digamsa Yantra. The page background reveals a ghost image of the full panorama, and on pages 62-63 you will see the full image.

To give another example, we look at the Delhi observatory from the upper surface of the west Jai Prakash Yantra. You might recognize the photograph from the cover of the book. It includes (l to r) a man in traditional garb standing on the upper surface of the Jai Prakash, a view of the gnomon and west quadrant of the Samrat Yantra with the Misra Yantra in the distance, the eastern Jai Prakash Yantra showing an entry to the lower level and steps to the rim, and

lastly, the north Rama Yantra. Each of these views is typical of how an ordinary camera might frame things, with an angle of view of about 60 degrees. The people shown in each of the views provide a clue about scale. As in the previous example, the views span all compass directions, beginning with a look toward the northwest at left, to the north in the second photo, to the east in the third, and ending looking southwest toward the Rama Yantra in the fourth photo.

Delhi

Spherical Panoramas from the Delhi Observatory

Rama Yantra, interior view of the
central column and radial sectors

Jai Prakash Yantra and Samrat Yantra
(above) and Rama Yantra (right)

Rama Yantra, steps to upper level of the Jai Prakash Yantra (west instrument),
Samrat Yantra, and Jai Prakash Yantra (east instrument)

Jai Prakash Yantra, with Misra Yantra and Samrat
Yantra (above left) and Rama Yantra (right)

Jai Prakash Yantra, interior

Jai Prakash Yantra, northwest entrance and
steps to the interior of the west instrument

Misra Yantra and Samrat Yantra (above)
and entrance to the Jai Prakash Yantra (right)

Samrat Yantra, top of the west quadrant

Samrat Yantra, top of the east quadrant

Jaipur

Spherical Panoramas from the Jaipur Observatory

Overview of the Jaipur Observatory
from atop one of the twelve Rasivalaya Yantras

Samrat Yantra, passage through the
gnomon (above) and west quadrant (right)

Samrat Yantra, east quadrant

Unnatamsha Yantra

Rama Yantra, interior view with
central column and radial sectors

Kapala Yantra. In the background above are the
small Samrat Yantra, Nadivalaya, Cakra Yantra,
Samrat Yantra, and Rasivalaya Yantras. To the right,
the Rama Yantra and Digamsa Yantra.

Jai Prakash Yantra from the rim of the north instrument. In the background are the Daksinottara Bhitti, Samrat Yantra, and Rasivalaya Yantras. Far right, small Samrat Yantra and Nadivalaya, .

Jai Prakash Yantra, looking towards the south instrument.
To the right can be seen the steps to ground level and the
passage to the north instrument.

Jai Prakash Yantra, interior, detail of nadir index

Digamsa Yantra

Ujjain

Spherical Panoramas from the Ujjain Observatory

Samrat Yantra, quadrants and
passage through gnomon

Digamsa Yantra, Nadivalaya, Daksinottara Bhitti,
and Samrat Yantra, from the top of the east quadrant

Digamsa Yantra

दिगंशा - यन्त्र

इस यन्त्र के बीच में गोल चबूतरे पर लगे लोहे के दण्ड में तुरीय यन्त्र लगाने पर ग्रह-नक्षत्रों के उन्नतांश (क्षितिज से उंचाई) और दिगंश (पूर्व पश्चिम दिशा के बिन्दु से क्षितिज वृत में कोणात्मक दूरी) मालूम होते है ।

तुरीय यन्त्र को इस प्रकार स्थिर कीजिये कि उसमें बने दो छेद्रो तथा ग्रह अथवा नक्षत्र का केन्द्र अपनी ऑंख से एक सीध में हो । दण्ड के सिरे पर लगे चक्र पर तुरीय यन्त्र की जगह होगा । दिगंश बतलाती है । वहाँ तुरीय यन्त्र पर लटकता हुआ डोरा यन्त्र के किनारे पर जिस जगह वाली सुई उन्नतांश होते हैं ।

अधीक्षक
ज्ञानकीय जौलोली ....

The observatory from the top of the Samrat Yantra

Varanasi

Spherical Panoramas from the Varanasi Observatory

Samrat Yantra, quadrants and
passage through the gnomon

View of the observatory from
the top of the Samrat Yantra

Nadivalaya, small Samrat Yantra,
and entrance to Digamsa Yantra

Cakra Yantra, Digamsa Yantra, Samrat Yantra, and Nadivalaya

Afterword

My first encounter with Jai Singh's observatories was in 1989, but the story behind my 30-year engagement with Jantar Mantar begins two decades earlier when I was an undergraduate student at Case Western Reserve University. I began my studies as a physics major, but had a strong leaning toward art and especially photography. I did a lot of growing up and soul searching in those years—the height of Viet Nam war protests and student activism—changing my major more than once and ultimately graduating as an independent scholar, with a visual thesis on the theme of *photography as language*. Photography became a way of life, both in a personal sense and as a means of earning a living. Concurrent with this, and perhaps most important, was a growing interest in the spiritual dimension of art. My first photography teachers, Nicholas Hlobeczy and Minor White, worked and taught from the point of view that heightened awareness and conscious attention were vital elements of artistic practice, and they referenced the writings of artists, psychologists, scientists, and teachers from the contemplative traditions in their teaching. I gravitated toward anything that touched on the spiritual in art, science, and philosophy, interested especially in perspectives that looked to the East for inspiration. It was not lost on me that the early 20th-century photographer Alfred Stieglitz and curator Ananda Coomaraswamy, good friends and both renowned in their fields, were highly cognizant of this deep connection between art and the traditions of contemplative practice.

Some 20 years later, for my first study-leave as an assistant professor at Cornell, I traveled to India and Nepal to photograph sites of sacred architecture. The venture brought me into close contact with extraordinary works of stone carving and remarkable

Left: Entrance to cave No. 34 at Ellora, India, 1989
Right: Seated Buddha, Ellora, India, 1989
From the exhibition *Made of Light: Photographs from India and Nepal*

spaces, including the Shore Temple at Mahabalipuram and the cave temples at Ellora, where I sometimes sat quietly for up to 30 minutes waiting for my film to expose in the dark recesses of these ancient places of religious devotion.

I worked in black and white in India, where the qualities of surface, texture, and light were most important. In Nepal, I was drawn to shrines and temples that were in everyday use, and worked in color to convey the vivid energies that I found there. The result was an exhibition titled *Made of Light: Photographs from India and Nepal,* which

Shivalaya Temple compound, Patan, Nepal, 1989. From the exhibition *Made of Light: Photographs from India and Nepal*

premiered at the MIT Museum in 1991 and traveled to the Sordoni Gallery at Wilkes University and the Berman Museum at Ursinus College between 1991 and 1992.

It was at the very beginning of this project, in February 1989, that I first saw the Jantar Mantar. I was in New Delhi to obtain letters of permission from the Archaeological Survey of India to photograph at the historic sites I had selected. Although the Jantar Mantar was not on my itinerary (it wasn't considered a work of sacred architecture) I decided to make a visit, as it would take several days for the permission letters to be ready.

Jantar Mantar observatory, New Delhi, India 1989

The first impression upon entering the observatory was that I had stepped into a different world. Inside the gate, the campus was spacious and surrounded by high walls and palm trees, and although it was in the midst of the busiest part of the city, there was a surprising sense of order and calm. The observatory structures were of course the dominant features, and unlike anything I had seen before.

I had a simple medium-format camera loaded with black-and-white film, but I spent most of my time walking about and looking, trying to fathom just what this place was. And then I began to see the beauty in the forms: their grand scale, and their precise geometry rendered in plaster and stone. The surfaces of the instruments were weathered, and marked here and there with graffiti, but this only added to the strength of the impression. As the enormity of what I was in front of began to dawn on me, I also began to make visual sense of it photographically. The elements of arcs, circles, triangles, cylinders, and steps, integral to Jai Singh's architecture of measurement, became through the camera's lens a language of abstract form. In the next few hours I exposed only three rolls of film, yet it was one of the most intense and concentrated work sessions of my career.

While the Jantar Mantar photographs were included in the exhibition of work from India and Nepal, I also felt that they needed to be seen on their own. A portfolio of eight photographs was published in the spring 1990 issue of *Parabola* magazine. That issue, with the theme of Time & Presence, included an interview with H.H. the Dalai Lama, and my portfolio immediately preceded an essay by Stephen Hawking entitled "Events in Time." The strong sense of abstraction and graphic qualities of the photographs also caught the eye of the architecture community, and a small portfolio was featured in the June 1992 issue of *Progressive Architecture.*

My time in India and Nepal was filled with vivid impressions and unforgettable experiences, but of all the subjects I had worked with, the Jantar Mantar held my interest most. It wasn't sacred architecture, yet it was compelling in its scale and the purity and elegance of its forms. It spoke to both my artistic and scientific sensibilities, and for years I entertained the idea of returning to make more photographs.

The opportunity finally came in 1999, a decade later, but it was only for a brief visit. I had been considering how to approach the observatories with a fresh perspective. At the time I had just begun to work with a 6x12 panoramic camera which took extremely wide

Jai Prakash Yantra at the Delhi observatory, top, and Digamsa Yantra, bottom, taken with a 6x12 panoramic camera, 1999

angle photos in a rectangular format similar to today's HDTV screens. The super-wide-angle camera was still very new to me, and it opened up visual space in a way that was completely unlike natural vision. This camera "saw" in a single frame far more than I could see without turning my head from side to side. So the photographs served mostly as sketches. I hadn't yet discovered how to use the new format expressively, and it took a year of experimenting with different subjects and ideas to learn how to use the "wide view" artistically.

In 2001 I returned to India for several weeks to do a more extensive photographic study of the Jantar Mantar. Earlier that year Apple had released Quick Time VR (virtual reality) technology, enabling spherical panoramas to be captured and viewed interactively on a computer. I instinctively felt this to be the way to extend my photographic work, and already had in mind an online virtual tour that would bring the observatories to a larger audience. So in addition to the 6x12 panoramic film camera, I also brought the camera and computer equipment for creating spherical panoramas. After a day of tests, it was clear that spherical panoramas surpassed any other photographic approach to convey the experience of being within these phenomenal spaces.

At the conclusion of the trip I had made approximately 40 spherical panoramas at Delhi and Jaipur, and over the next year I built a website to share my experience of the observatories with others. At the same time, I began rendering the spherical panoramas as large prints and presenting them in group and solo gallery exhibitions.

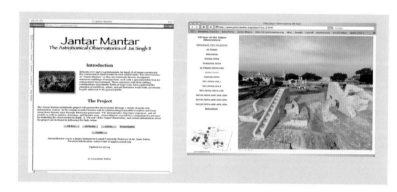

www.jantarmantar.org, launched in 2003, included a virtual tour of the Jaipur observatory

The original website was well received, and I soon saw the need to expand it to create a "virtual museum" about Jai Singh's observatories. I returned to India in 2004 to do additional photography at Delhi and Jaipur, and to add the observatories at Ujjain and Varanasi. This enabled me to create a comprehensive virtual tour that included each of the four existing sites. Over the next decade, additional research and the creation of 3D models by talented architecture students allowed me to craft explanations and resources for online visitors to better understand the observatories and how they worked. Interest from around the world has kept me busy learning and developing more ways to explain and present them. These have included gallery exhibitions, online astronomy applications, and interactive presentations in planetarium theaters and science centers.

Throughout this 30-year process, the very interdisciplinary nature of Jai Singh's observatories, the way they incorporate astronomy,

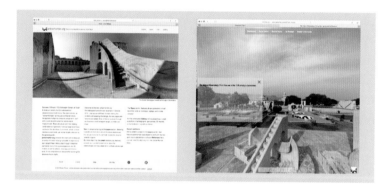

www.jantarmantar.org as of 2019. The current website includes virtual tours of all four observatory sites, an expanded educational area, galleries, and multimedia components.

astrology, mathematics, art, and architecture, has been an inspiration. Although viewing the sky with the naked eye no longer has the scientific significance it did before the telescope, it will always be a part of human experience—our way to remain connected with the immensity and mystery of the world in which we live.

First Encounter

New Delhi, 1989

Suddenly I was in a different world, a world of architectural fantasy alive with complex forms—curves, arches, cylinders, triangles, and steps everywhere—and on a scale so monumental one felt, even without understanding why, an authority behind the forms.

Here was a work of architecture, or perhaps one could say sculpture, that carried precision in its form and mystery in its effect. Powerful forms, with dignity, that spoke of the circular, cyclical movement of time and were lyrical, poetic.

Here, just a few hundred feet from one of the busiest parts of
New Delhi, was a world apart, and I had fallen under its spell.

Jantar Mantar has been an inspiration to artists, architects, poets, and countless others in all walks of life. The sculptor and designer Isamu Noguchi photographed the observatories in 1949, and Le Corbusier drew upon the unique shapes of the Jantar Mantar in his design of the capital buildings at Chandigarh. The Spanish poet Julio Cortazar photographed the Jantar Mantar in 1968 and combined those images with a lengthy muse on existence in *From the Observatory.* A scene in the movie *The Fall* (2006) was filmed at the Jaipur observatory.

For me, Jantar Mantar was the beginning of a creative project that has spanned almost three decades. Jai Singh's observatories have made a lasting impression, and this book is just one of the ways I take note, and give thanks.

Index

Acknowledgments

The author wishes to thank the following for their contributions and assistance.

Bei Xu created the majority of the 3D models and graphics for jantarmantar.org and this book.

Robin Liu created 3D models for the website and produced renderings and drawings that were used in early versions of this book.

Dr. Nandivada Rathnasree, Director of the Nehru Planetarium in New Delhi, was invaluable in assisting me with the photo shoots in 2004, and has been a constant support and advocate for my work with the Jantar Mantar.

Dan Neafus and KaChun Yu of the Gates Planetarium at the Denver Museum of Nature and Science recognized the potential for presenting the Jantar Mantar in a planetarium dome and opened the door to my exploration of immersive space.

Mark Subbarao, Director of the Space Visualization Laboratory at the Adler Planetarium, another early supporter and an ongoing collaborator, taught me many things both astronomical and immersive. The theme of immersion that runs through this book emerged during one of our work sessions at the Adler.

Photo Credits

All photographs in this book are by the author except as credited within the text or noted below.

Photographs on pages xii, xiii, and xv are by Anisha Shekhar Mukherjee.

The photograph on page xvi is by Snehanshu Mukherjee.

The photograph of the author on this page is by Anna Salamone.

About the Author

Barry Perlus is an Associate Professor and former associate dean in the College of Architecture, Art, and Planning at Cornell University, where he has been teaching courses in photography at the graduate and undergraduate level since 1984. He received an MFA from Ohio University, and BA from Case Western Reserve University.

With an avid interest in both art and science, his artistic practice includes projects in photography and digital media, notably panoramic and immersive imagery. As an artist/scholar/author/educator, Professor Perlus has received numerous grants to advance his work, including from the Mario Einaudi Center for International Studies at Cornell and the Graham Foundation for Advanced Study in the Fine Arts. Portfolios of his photographs have appeared in national publications such as *Parabola* magazine and *Progressive Architecture,* and his work has been shown in more than 50 one-person and group exhibitions both in the U.S. and abroad.

Professor Perlus's teaching, research, and creative work have been recognized with many distinguished awards, and most recently he was named the 2015-2016 Webster Lecturer in Archeoastronomy by the Archaeological Institute of America.

He resides with his wife Anna in the Finger Lakes region of upstate New York. There, in addition to photographing the natural world around him, he can be seen enjoying expansive views of the night sky, tending his orchard, or hiking the Black Diamond Trail.

This book is dedicated to Anna, my dear companion in all things creative and source of support and encouragement at every step and turn this project has taken.